BUILDING

WITH

HEMP

STEVE ALLIN

SEED PRESS

Published by SEED PRESS
Rusheens, Kenmare,
Co. Kerry, Ireland.
seedpress@eircom.net

ISBN 0-9551109-0-4

The details and information are provided in good faith and are believed to be correct at the time of writing. Please let me know of any amendments or additions which you think may be useful for future editions.

Any building projects undertaken as a result of reading this book are done so at the readers own risk. As circumstances, materials and workmanship cannot be guaranteed by the author.

I have endevoured to credit all photographs and information within this book to the owners of such items. If I have failed in this way please let me know and I will make amends in the next edition.

Printed on 100% tree-free paper

Printed & bound by Reggiani SPA in Italy 2005

THANKS

It is not until you write a book yourself that you understand why the thanks section in others are as big as they are, and like anybody writing or compiling a book of any sort, I did not do this one on my own. From the start of my interest in hemp building I have had the unfailing support of my wife Aia, not only for her constant encouragment, but also as she speaks fluent French she was able to translate all the early literature and brochures produced by the first hemp building suppliers and associations in France. She also did some of the better illustrations in the book and is an exellent proof reader.

Other members of my family have also helped out, my mother Betty with research and newspaper clippings, my mother-in-law Loretta Leu with translations and photographs, Ajja S.F. Leu for helping with his fantastic computer graphics skills and the rest of the Leu family for their encouragement.

Throughout the development of my understanding of how the whole thing worked I have been helped with the architectural expertise of longtime friend Anthony Cohu and with the scientifically knowledgable minds of my friends Anne & Andy Garrett and Redmond Roche. A big part of comprehending the subject was the understanding of how lime works, in this I have been helped and supported by Edward Byrne of the Traditional Lime Co.

Apart from my own experience much practical know-how has been acquired by visiting France, to see first-hand and photograph projects there. In this I have received wonderful assitance from Hélène Denis, Charles Coignard, Jackie, Alain Dupays, Catherine et Claude Eichwald and especially Laurent Goudet for his time and pictures. I have been helped gaining experience and having my own validated by the members of Construire en Chanvre and the many self-builders who have shared their experiences with me and let me into their homes to take pictures of their handiwork. For this I have to thank; Jeff & Cathy, Brian & Patricia, Lorian & Adam, Pascal Petit, Sebastian & Róisin, Olivier & Maia, Maghda O'Mahony, Bas & Ester, Fatima,Lia,Jules & Alba, John & Katie, Donal & Cathy, Alex Schmeider, Robert & Paul, Chris & Paula, Merrick, Dominic & Ina, Dave, Thomas & Ulrika.

Additional thanks to Yvan Saint-Jours, Lucy Farmer, Ivan Watkins, Tim Yates of the BRE Ltd. Bruno Mandelli & Yves Betrencourt at LCDA. Kevin & Trea for the website. Susan McCarthy at Hemp Cottage for hemp products, Arthur Hanks of the Canadian Hemp Trade Alliance, Oliver Holbien Pete & Olga Ireson, Richard @ Sinhouse, Terry & Sarah, and my good friends Dean & Deirdre.

Finally I must thank Tessa Dennison who has helped an amateur layout artist like myself to achieve the look and presentation of this book as it is.

The author, out standing in his field.
(Sorry I couldn't resist it).

Foreword

Before the reader gets immersed in this fascinating and groundbreaking primer on building with hemp, it is worth remembering the context in which all building activities now take place.

Globally, the construction industry consumes 50% of all natural resources, a destructive and unsustainable trend epitomised by rampant urbanisation. In recognition of this stark reality, the Earth Summits of Rio (1992) and Johannesburg (2002) identified a way forward for the global community - sustainable development - based on thirteen key principles where:

- Resources are used efficiently and waste is minimised by closing cycles;
- Pollution is limited to levels which natural systems can process without damage;
- Natural diversity is valued and protected;
- Local needs are met locally where possible;
- Good food, water, shelter and fuel are available to all at a reasonable cost;
- Satisfying work in a diverse economy is available to all;
- Good health of the community is protected;
- Environment is not damaged by access to facilities, services, goods and other people;
- Personal safety of the community is secured from crime and violence;
- Skills, knowledge and information are accesible to all;
- Participation in descision making is extended to the whole community;
- Culture, leisure and recreation opportunities are available to all;
- Local distinctiveness and character are valued and protected;

Those of us involved with building, whether amateur or professional, designer or builder, have a duty of care not only to the larger global environment, but the local community where we live and work. We can each contribute to the effort to create healthy, affordable and practical building solutions that respect the natural systems on which we depend, by recognising and responding to these principles.

The author and myself have each been fortunate enough to live for the past thirty years in one of the least polluted and industrialised parts of Eurpoe - Ireland, and one of the most beautiful parts - the coastal valley system of the South West. This corner of Ireland has the wettest climate in Western Europe, where a sunny summer's day can still have a relative humidity of 70%. Add to this wind-driven salt spray that can reach up to 20km inland, and it is clear why certain building technologies have been developed and others dropped. It is a real challenge to design and build comfortable and durable buildings in such an environment - a stern test of building methods, both traditional and innovative.

Vernacular building used local materials such as earth, stone, wood, thatch, and lime mortar until the end of WWII when the availability of cement enabled the first mass-concrete walls to be built. In turn, concrete blocks and tiles came to be the prevalent building technology, and are now considered as standard construction. Timber, the best known of all the crops used in building, has only recently become a significant construction material in Ireland with the result that 10% of all newly-completed dwellings now use timber-frame construction,albeit with a skin of cement-rendered blockwork. The vernacular tradition has also been based on self-building, a practice which continues to this day with over 50% of new owner/occupied housing in Ireland self-built. The author and myself have each designed and built our own energy-efficient houses, using many of the techniques and materials described in this book, as well as doing so for others who are dedicated to creating a new vernacular appropriate to current lifestyles.

As a country where the agriculture sector is still a significant producer, Ireland is well placed to grow other building crops besides timber, and create new materials and products.

Hemp is the most promising of these, currently enjoying a renaissance in many European countries as an ideal crop with a wide range of useful properties. This book reviews both the theoretical background and practical applications of building with hemp, which can compare with conventional materials in terms of resistance to moisture, structure and durability, yet give superior thermal performance and fire resistance. Its production could become very important to 21st century European agriculture, together with the production of biofuel crops, such diversification is essential for the survival of agriculture in Europe, and the long-term sustainability of the global environment.

Anthony Cohu R.I.B.A. past chairman of Cork Environmental Forum

CONTENTS

Opposite; Hemp house in Montholier, France. Built as part of a test of hemp as a building material compared to strawbale.
Previous; Pierrefleur "Terrain d'Adventure" Lausanne, Switzerland.

IN this book I will describe how to build using materials derived from the hemp plant, however I must get certain things clear from the outset. You cannot smoke the house or see the joints in the walls. These comments are always made when first mentioning the subject and are therefore best to get out of the way at the beginning. As for consuming the psychoactive element of the plant whilst building, I will leave that up to the readers discretion but add that site safety should be observed at all times!

Having got that out of the way, what is building with hemp all about and why would anyone think of using a material such as this to build a house or any other building with? For many years now architects and builders have been searching for alternative systems and materials to achieve lower environmental impact and healthier and more economic solutions to the basic human need for shelter. Some of these ideas have come from adaptations of ancient and traditional techniques, others have grown from various uses of modern materials whether they were intended as building materials or not.

The resurgence in interest in techniques such as cob walls made of earth and straw or log or cordwood cabins reflects a desire to get back to more simple craftsmanship using materials immediately at hand. Other materials such as strawbales and used car tyres have been seen as a waste material put to more positive uses in newly evolved methods.

In other areas of construction, traditional skills and methods are making a comeback due to the wish to conserve old houses with the correct materials both for the sake of appearance and the way in which the materials work together. Trades such as thatching, stonemasonry and the use of lime mortars and plasters need a certain level of skill and knowledge in order to apply them correctly, but training in these fields is now available again and in some cases are grant funded by governments to ensure their survival for the future.

In the past houses were mainly designed and built by tradesmen who had served long apprenticeships using methods and designs evolved slowly over centuries. In more recent times the profit motive and the desire to rush through the job and let the insurance companies sort out any flaws have led to attempts to produce systems suitable for factory made components and relatively unskilled labour. This provides homes with little or no character.

I first learnt about hemp building whilst reading the now famous book " THE EMPEROR WEARS NO CLOTHES" by Jack Herer. This book outlined the many uses, past and potential, of the plant and the probable reasons for its' prohibition (for more information go to page 16 & 17). There, in the appendix, was an extract from the "Isochanvre" brochure. As I was halfway through building my own house at the time I decided to take a small gamble and import some of the product and use it as a plaster.

The results, once I had mastered the technique, were successful enough that I had soon sold the remainder and was ordering more for other projects.

It quickly became apparent that wherever in the world one is, it would be better to be producing the hemp nearer to home and so for many years after this I have been attempting to start a hemp industry here in Ireland. However, it can be a bit like the chicken and egg syndrome in that one needs a market big enough to justify the expense of putting the processing machinery in place. To that end I have now been concentrating on communicating the details of this revolutionary method of building and, of course, this is the primary reason for writing this book.

To sum up, what I have discovered about building with hemp is that it is possible to create an "ECO" home that can look like any style of building. One that can be used in the same way as a normal house but have all the properties of sustainability and environmental impact one can reasonably expect at this point in time. These aims are now more clearly understood and all materials and methods are now being scrutinised for their energy consumption and overall impact. All manufacturers of building products now eagerly promote such " green " aspects of their products.

So here is a book in which I hope you will find all you need to know as to why and how to design and build your home using the products of the hemp plant, plus of course a few other ingredients. As this is the first book on the subject (which is itself in its' infancy) it will of course, need to be updated when more projects have been completed, but for now I hope this is enough to inspire you in your endeavours of building for yourself or others an ecologically sustainable home that is healthy and a joy to live in.

All the best
Steve Allin.

Ireland, 2005.

Bast Fibre Crops

The bark of plants which contain fibres that can be utilised for cordage or matting are called bast fibres and so as a group are classified as bast fibre crops.

In this group are many species of plants, the most common being flax and hemp which have, from the earliest of times, been used to make fine linens. The stinging nettle also has been used in this way and recently clothes made of ramie from the east of Asia have been seen on the racks of clothing stores.

Similar plants producing coarser fibres such as jute, which is mainly used for carpet and sacking, and kenaf used mainly in paper manufacture, are also crops with similar constituents of an outer bast containing long fibres and an inner hollow woody core.

Crops such as sunflowers produce much waste material after the seed has been removed and this has recently been proposed as a possible material to be used in a.similar way to hemp hurds in building

BAST FIBRE CROPS;

JUTE; *Corchorus capsularis.*
Primarily grown in Bangladesh and India also in China, Thailand and Mayanmar. Fibres coarser than hemp.

FLAX; *Linum usititissimum.*
Formerly grown extensively in Ireland, Belgium and Germany is primarily produced now in Turkey and Russia.

RAMIE; *Boehmonia nivea.*
Grown in China, Brazil, Philippines, India, South Korea and Thailand. Perennial crop can be harvested up to six times a year. Fibre not as durable as other textile fibre. Needs chemical degumming process.

KENAF; *Hibiscus cannabin*
Native of tropical Africa now cultivated in India and Southeastern Europe. Also called Ambari hemp.

NETTLE; *Urtica dioica.*
Common to Europe and Asia, is usually seen growing on waste ground, used to be used as a substitute for flax, but the lack of a valuable seed for oil meant was uneconomic as it needed the same rich land and the same quantity of labour.

Left Top; Peeled hemp stem.
Left Middle top; Jute thread.
Left Middle bottom; Ramie wood chips.
Left Bottom; Stinging nettles.
Left; Sunflower seed crop

Linen

Linen has recently made a comeback as a fabric of choice for clothing, as it feels cooler than cotton and is much more hard wearing. Once used for all domestic coverings such as bedsheets and tablecloths, it has largely been replaced by cotton and polyester.

Despite this, all such goods are still referred to as bed linen or table linen. As anyone with experience will tell you, for comfort and appearence the modern materials don't come near traditional linen, especially when it comes to the cool crisp comfort of linen bedsheets.

However, the labour intensive manufacture of long fibre threads compared to the incredible speed with which short fibres such as cotton and polyester can be spun has meant that these materials have become more of a luxury item.

The manufacture of linen from flax goes back to ancient egyptian times where the mummies of pharoahs where wrapped in fine linen strips and the skill of spinning and weaving were highly valued. In early European civilisations a lady of means would have brought spinsters (young unmarried women who spent all their time spinning) with her as part of her dowry. It is thought by some that the lost arms of Venus de Milas were originally in the position of spinning with a whorl spindle. Linen production reached its peak in the age of water power especially in the northern province of the island of Ireland called Ulster. There the new potential of watermills were put to use powering the machinery developed for all the stages of flax processing.

The last period of large scale linen production was during World War I & II when it was in great demand as the strong lightweight covering of wooden aircraft fuselages. Nearly all linen production is now carried out in countries where labour is cheap, as despite the level of mechanization to the processing of linen fibre into thread, it is still a slow process.

Right; Flax seed heads from which linseed are extracted.
Below; Shop display of linen goods.

HEMP

Renaissance

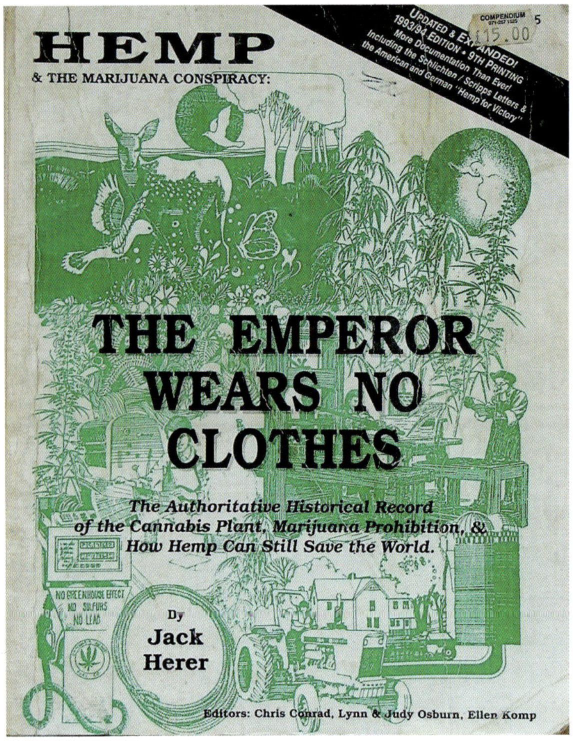

Back in 1985 a book was published that was to have far reaching implications and the major facts within be quoted and repeated by people all over the world.

Written by a marijuana activist by the name of Jack Herer, the main premise of the book was that cannabis was classified a a dangerous narcotic drug to protect the financial interests of a few industrialists and bankers.

The book included vast amounts of information about the history of marijuana use and as in "The marijuana farmers " by Jack Frasier, published years earlier, documented the early hemp farming endevours of the founding fathers of America .

Herer and his co-editors brought together many aspects to the story leading up to the passing of the Marijuana Tax Act which had the effect of outlawing what had been for centuries a widespread agricultural activity, supplying a very necessary commodity.

Although elements of the prohibition story added considerable weight to the argument for legalization, the main result of its' widely disemminated information especially by the new medium of the Internet, was the possibilities that hemp offered those looking for sustainable solutions to the global problems caused by the polluting petrochemical industries.

Top; The cover of the "Emperor wears no clothes".
Left; A selection of some of the many publications featuring the hemp issue.

The potential of hemp to supply cellulose to manufacture paper and reduce the need to use trees, or rayon to replace nylon. The ability of hemp to provide biomass for fuel was presented alongside information on the food value of the seeds and the potential of the medicinal use of the TetraHydra Cannibol, the contraversial drug contained in the resin of the plant.

Going even further they issued a 10,000 dollar challenge to anyone who could prove them wrong on the claims made in the book; that hemp was the only renewable natural resource able to provide the overall majority of our paper textiles and food, meet all the worlds transportation, home and industrial energy needs, reduce pollution, rebuild the soil and clean up the atmosphere. As far as I know, no-one has made a successful bid to claim that prize. Since then Hempflax in the Netherlands heve increased that challenge to 100,000 euros.

It is somehow ironic, but fitting, that a plant whose most notorious element, its mildly psychedelic resin which had been so popular with those sympathetic to a more caring approach to the planet, should then be discovered to hold many of the keys to turning back the tide of destructive greed.

Hemp Products

Since the publication of "The Emperor Wears No Clothes" a vast industry has grown up to supply at least some of the possibilities that were put forward. Of these, food and clothing have seen the greatest success.

At first existing hemp fabrics, mainly lightweight canvases, were used to make hemp jeans, shirts and jackets. However to satisfy a more discerning consumer who wanted garments as comfortable as those they were already used to, improvements had to be made. So with perseverence and refinements to both the textile manufacture and the design of the garments it is now possible to purchase clothing made from knitted or woven textiles using blends of hemp with cotton, wool, camel hair or silk.

The addition of hemp gives many beneficial qualities to the wearer. As with linen made from flax, hemp can be susceptible to creasing, but blended with other fibres qualities of better temperature regulation (cool in summer, warm in winter) absorbtion of perspiration, and in the case of T-shirts better resistance to ultraviolet radiation, can be appreciated by those wearing hemp clothes.

On the food front the nutritional value of the combination of omega 3 essential fatty acids in the form of linolic and linalonic acids and the proportions in which they are found in the oil expressed from the seeds have been widely accepted. Hemp oil is now readily available and is especially recommended for those who need to dramatically improve their immune system.

The seeds or nuts are incorporated into all kinds of foods from chocolates to pasta and can be sprouted to be eaten raw in salads.

The oil is also used in non-food products such as the range of toiletries marketed by "The Body Shop".

A further use, not only of the seeds, but also for the flower heads, is as a replacement for hops as a flavouring for beer which although sounds like a novelty is in fact quite delicious.

Below left; Tube of hemp oil hand protector cream from 'The Body Shop' ideal for use when building with hemp.
Below right; Collection of hemp goods now available.

History

Above; FRANÇOIS RABELAIS
Below; Medieval Garden in Mallow, Co Cork, Ireland. Featuring Hemp Hut built in 1999 by author.
Right; Decorated column in German church.

Hemp has been used by man for thousands of years. In fact some anthropologists believe the eating of the seeds by early man led to an unintentional ingestion of the psychoactive element present in the resins coating the calyx which surrounds the seed.

This is thought to have helped mankinds contemplative abilities and encouraged the brain to develop in the way it has today. Whether this is true or not, the plant seems to have been associated with humans from very early on in our history.

Numerous artifacts have been found by archeologists made from the fibres; from simple woven belts and shoes of the ' First Peoples" in America and China, to fine hemp linens discovered in Japan. The medicinal qualities of the plant have also been mentioned in many of the early herbals. The earliest being the Chinese emperor Shen Nung, but Dioscorides, Herodatas and Culpepper have also all detailed its uses.

But perhaps the most detailed account of the many qualities of the plant were written as an allegory by the French writer François Rabelais.

In his story "Gargantua and Pantagruel" a parody of epic tales, Rabelais writes in extreme detail of life in early sixteenth century France. In this story of two giants, Rabelais takes many swipes at church and state and many other groups besides, but within the phantasmagorical text are many descriptions of late medieval life. At the end of book three in the work, he tells us about the amazing herb pantagruelion. This is obviously hemp and although he doesn't mention the mixing of hemp with lime, he does write "Seldom is the mortar, lime, or plaster brought to the workhouse without it". This is very likely a reference to lime mortar used in buildings that would have been brought to the site in hemp sacks, used no doubt for their great strength and the ability to absorb a good deal of water needed to keep the lime materials damp.

Prohibition

To prohibit the growing of a plant is a pretty extreme measure, and yet that is what has happened with cannabis or hemp.

To understand some of the reasons for this predjudice we must go back to the late Roman times of the emperor Theodosius, who in 375 CE outlawed the practice of all religions other than Christianity. This would have included many old beliefs whose adherents often sought illumination through the use of intoxicating incense. Many ancient texts from diverse religions talk of illumination reached by means of entering an altered state and this seems to have, in some way, threatened the influence of the christian rulers.

Today it is those same irrational fears that are evoked in the style and method of many media reports about cannabis that mirror the period of the witchtrials.

When Harry Anslinger was appointed as head of the Federal Narcotics Control Board in 1932, he can hardly of anticipated his future success. Originally controlled as pharmacuetical medicines began to emerge, the definition "Narcotic" seemed to speak of the fear sung of in the jazz lyrics "The stuff that dreams are made of, the thing that white folks are afraid of". To be able to convince state after state, nation after nation, that there is a reason to prohibit the growing, trade or use of an intoxicating plant is some feat.

The result today of this completely unsuccessful prohibition is a black market drug trade worth an estimated 7000 billion dollars !!! With a cost to the American tax payer of 50 billion dollars a year!

Below; The famous "FABULOUS FURRY FREAK BROTHERS "drawn by Gilbert Shelton.

Global

Much has been written about the potential of hemp to supply many of the commodities we use across the world today. Of course to be able to do that many hundreds of thousands of hectares of land would have to be put to use growing the hemp crops necessary to realise such a plan.

Despite food shortages in some areas of the planet, there are many regions of the world, especially in western Europe, that have the type of land suitable for the cultivation of hemp that is at present lying idle. This is as a result of both economics and global trade agreements.

With the predicted upcoming oil shortages, the need for more sustainable methods of food production will hopefully lead to a return to more appropriate methods of agriculture, together with developments in utilising non-food crops.

It is not possible to change overnight these methods, but the possibilities are being investigated and demonstrated in regions that have always grown crops such as hemp or are now doing so again.

Countries such as China and France have never lost their hemp industries, and places such as Russia and other areas of eastern Europe have the remnants of such facilities. In the 'West' countries such as Canada, England, Spain, Germany and the Netherlands have begun hemp processing once again. There are also attempts being made in thc Czech Republic, Australia, India Ireland, Finland, South and North Korea and South Africa.

Below; Map of the world showing major centres of hemp processing (red) and areas where hemp could be grown (green).

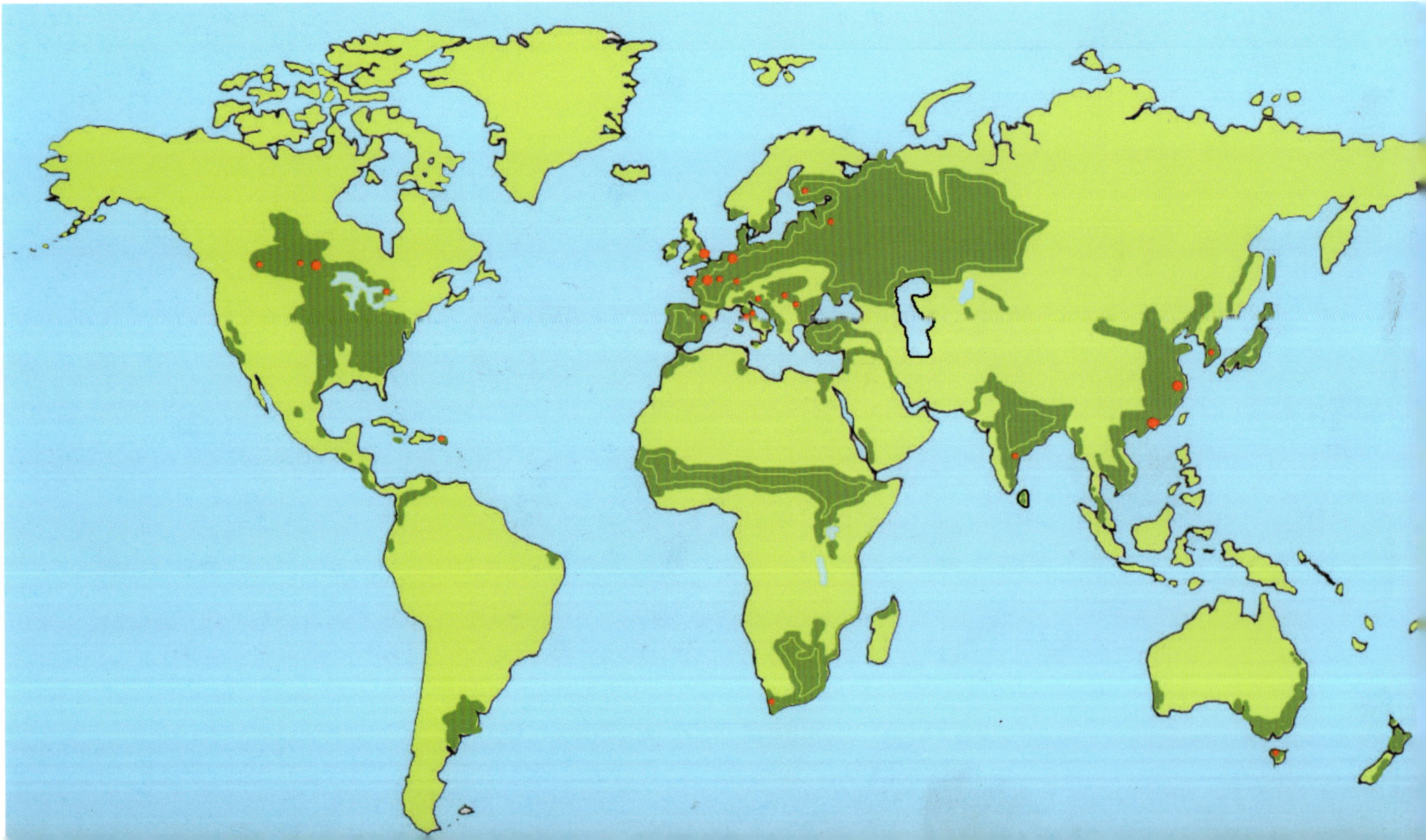

The technique of hemp building is now beginning to spread around the world. Naturally as it is a concept that originated in France, in that country the pratice is now widespread.

Projects utilising hemp as a building material are regularly featured in magazines such as "La Maison Ecologique" and centres of alternative architecture often offer training courses in its use.

In Britain and Ireland hemp building has been adopted by some, with several hemp buildings in England, mainly due to the activities of architect Ralph Carpenter, and in Ireland where it has become popular as a plastercoat used in the renovation of old stone cottages.

In Switzerland the use of hemp with lime and gypsum has been adopted as a result of contact with the system promoted by Yves Khün.

In all regions where there is a hemp industry, there is now a growing interest in developing such a use and I have reports of tentative beginnings in Canada, Belgium, South Africa and the Czech Republic.

In Germany and Austria the building trade has been using hemp fibre insulation for some years and this is now also being used in France, both in manufactured panels and as loose fibre.

Left; Hemp plaster being applied to an old house in the Czech Republic.

Botany

The leaf of cannabis sativa, with its palmate form and long lanceolate, serrated leaflets, is probably one of the most commonly seen icons of modern times.

On the covers of mobile phones, emblazoned on T shirts, embroidered on hats, in silver as an earring or as a logo on a cosmetic, it crops up everywhere.

The genus; Cannabis is divided into three subspecies:

(i)Sativa; all hemps and some marijuana, literally, seed sowing.

(ii)Indica; originates in himalayas especially India, psychoative.

(iii)Ruderalis; originates steppes of russia, programmed to flower after certain number of days as adaption to short growing season.

An annual plant growing up to 10-18 feet if encouraged to do so but usually only reaching 6-7 feet as a hemp crop when grown close together. The leaves grow opposite each other on a hollow stem. Fine fibrous roots branch off the central woody tap root which anchors the plant as it shoots to the sky. As the plant nears maturity the fibres in the bast strengthen and the woody core hardens to prepare to support the ripening seed head.

These seeds are more accurately described as nuts being a kernal in a shell, they are very high in essential fatty acids and are fed on voraciously by small songbirds, which can be seen in great flocks at seed harvest time.

Top left; Hemp seed
Middle left; Hemp fibres
Left; Hemp hurds
Right; The famous leaf

Agriculture

The manner in which hemp is grown dictates which of the materials the hemp plant can provide.

To obtain long strong fibres it is sown in rows 100 cm (4 ins.) apart, in order to promote vigourous growth straight upwards as the plants race with each other for the light. The more densely it is sown in these rows, the shorter and finer the stems are and therefore also the fibres.

In most climates hemp is sown in april to miss the last frosts, but it can be sown as soon as early march if conditions are right. The young plants are emerging from the ground after the first week and the growth so vigorous in the 2nd and 3rd month as to achieve 100-300 mm a week.

When hemp is grown in this manner the preparation of the soil is of paramount importance, for if this is done correctly then nothing more need be done until harvest time. It is very important to thoroughly till the soil in order to ensure even germination and to roll the ground after the seed is sown to ensure that it is possible to harvest as close to the ground as possible and not damage the harvesting machinery.

When hemp is grown for seed it is grown as individual plants which are planted 0.5-1 metre apart, to allow the branches that sprout from the leaf nodes to grow out into the light and fully develop their seed heads.

Top; Pneumatic seed drill used for accurate sowing of hemp for fibre.
Right; Closely planted fibre hemp.
Far right; Hemp grown for seed.

Harvesting

Depending on what business you are making from your hemp crop, the time for harvesting will either be as the seed begins to set (for fine soft fibre) or when the majority of the seed is mature (for the dual harvest of seed and fibre).

In regions where hemp production has been continuous older types of harvesting machinery are still in use. In countries that have restarted the industry, faster more modern methods have been utilised, especially disc and drum mowers that can race across the fields leaving the stalks in sways or rows for an adapted round baler to pick up and bale.

The traditional method of leaving the stalks on the ground to "rett" can result in an uneven condition of the material and so, in countries where the weather at this time of year is unpredictable, it is better to remove the foliage before cutting, so the extra water content of the leaves does not cause problems. This can be done mechanically, either collecting the foliage for fragrance extraction or the leaf matter can be returned to the soil to decompose naturally.

Top; Round bale of organic hemp.
Bottom; "Grasshopper" drum-mower.
Right; Mechanically defoliated crop.

Processors

As with all processing of agricultural crops the separating or decorticating (the removal of the cortex or skin) of hemp was done by hand until the machinery was invented.

However, as hemp has the longest strongest fibres of the plant kingdom this was not an easy task. Most machinery involves rotating componants and as a result provides many opportunities for those fibres to wrap around them and cause havoc. To produce fibres from hemp that are usable in the spinning industry entails slow movement through the machinery to avoid such problems, and as a result is largely uneconomic in the West. This means that with faster processing the resulting fibres are usually produced in an unaligned form, suitable only for the production of matting used as the reinforcement of panels for the automotive industry as insulation in the construction industry and to produce speciality paper.

The largest hemp processors in Europe are in England, France and the Netherlands, others operate in Austria, Italy, Spain and Switzerland.

Of these, La Chanvrière de l'Aube, east of Paris, has been operating the longest. This co-operative has been in operation continuously since 1973 and currently processes hemp from 6,000 hectares.

Top; Organic hemp processor in Brittany.
Middle & Bottom Right; hemp packaged for building use.
Below; L.C.D.A. factory in Bar-sur-Aube, France.

Rotation system

Hemp can play a vital role in organic agriculture. For the organic system to work efficiently weed control is essential for maximizing yields, if we are to expect growers to provide us with an ecological alternative to the food stuffs produced with chemical fertilizers and herbicides.

Systems must be established that replace these chemical solutions and prevent the need for genetic modification. This entails devising rotation systems that leave the soil in the right condition for the following crop on a year by year basis. Hemp provides a great deal of shading of the ground as it grows, and when harvested, the land is left with fewer weed seeds to germinate the following year. The leaf debris, falling from the plant as it grows, puts back as much as 40% of the nutrients used by the hemp during growth.

In such a rotation system fibre hemp can be grown after a nitrogen fixing, green fertilizing crop such as clover or alfalfa. It is best grown before winter sown crops such as wheat which benefits from the weed-free seedbed provided by the previous hemp crop with improved yields of between 10-20% !!

Right top; Shade beneath canopy of hemp.
Right bottom; Weed free ground beneath mature seed crop.
Below; Hemp field

Environment

One of the aspects of using hemp products that is constantly put forward is that it is better for the environment.

The primary reason for this could be said to be, that with building materials we are replacing a material which the production or, in the case of gravel, the extraction of, is unsustainable.

From an agricultural point of view hemp is also beneficial to the environment even if it is not grown under an organic system. The fact that as a crop, it needs no chemical herbicides or insecticides is an obvious advantage. Less understood is the way in which, by shading the ground and therefore keeping it cool, it is less likely to dry out and suffer from erosion by wind or sudden rainfall.

During the growth cycle much of the leaf matter is returned to the soil. As the plant grows taller the leaves are shed when they are no longer needed by the plant as they become shaded. If the crop is mechanically defoliated all such vegetation can be returned to the soil in a shredded state which decomposes quickly.

During the harvesting period small proportions of the hurd are left behind no matter how efficient the gathering technique, though the stems are cut as near to the ground as possible to maximise the yield. Unlike flax which is pulled root and all, a substantial portion of a hemp root is left in the ground which acts not only to hold the soil together, but also acts as a pipe to aerate the soil.

If the hemp is being grown as a dual purpose crop and the seed is to be gathered as well, there is the added advantage of the increase in songbird populations and the rich manure they deposit as they feed on the seed before the farmer gathers it in.

Other low environmental impact aspects of hemp would be associated with reduced energy input and non-toxic additives during the manufacture of hemp products.

Above; Hemp tap root.
Left; A field of winter wheat

Hurds

HEMP HURD ANALYSIS	
CELLULOSE;	37 %
HEMICELLULOSE;	16.5 %
LIGNIN;	21.8 %
PECTIN;	5 %
EXTRACTIVES	8.9 %
ASH ,SILICA, TANNIN + OTHER	10.8 %

When it comes to building with hemp, it is the wood-like core of the stem we are primarily interested in.

The quality, form and size of the particles that this material is produced in will be important when used as an ingredient in the building material. From both the perspective of the farmer and the builder the crop is better harvested when mature, as the yield is higher and the hurds firmer and so able to withstand greater compression. The size and proportions of the particles will be determined by the type of machinery used to decorticate the hemp stems, the longer particles giving greater lateral strength, the smaller ones a denser material with (in the case of plaster a smoother surface).

Other considerations when selecting hemp hurd for building are that it should be damaged, from retting, as little as possible. Contamination with leaf matter and foreign material should also be at a minimum, of course. There have been cases of customs and excise officers in certain countries refusing shipments of hurd on the basis that it contained enough leaf material to be classified as illegal marijuana, even though they had acknowledged that it was grown from legal low THC seed!! Absurd but true.

Left; Cross section of hemp stem
Below; Diagram of componants of stem

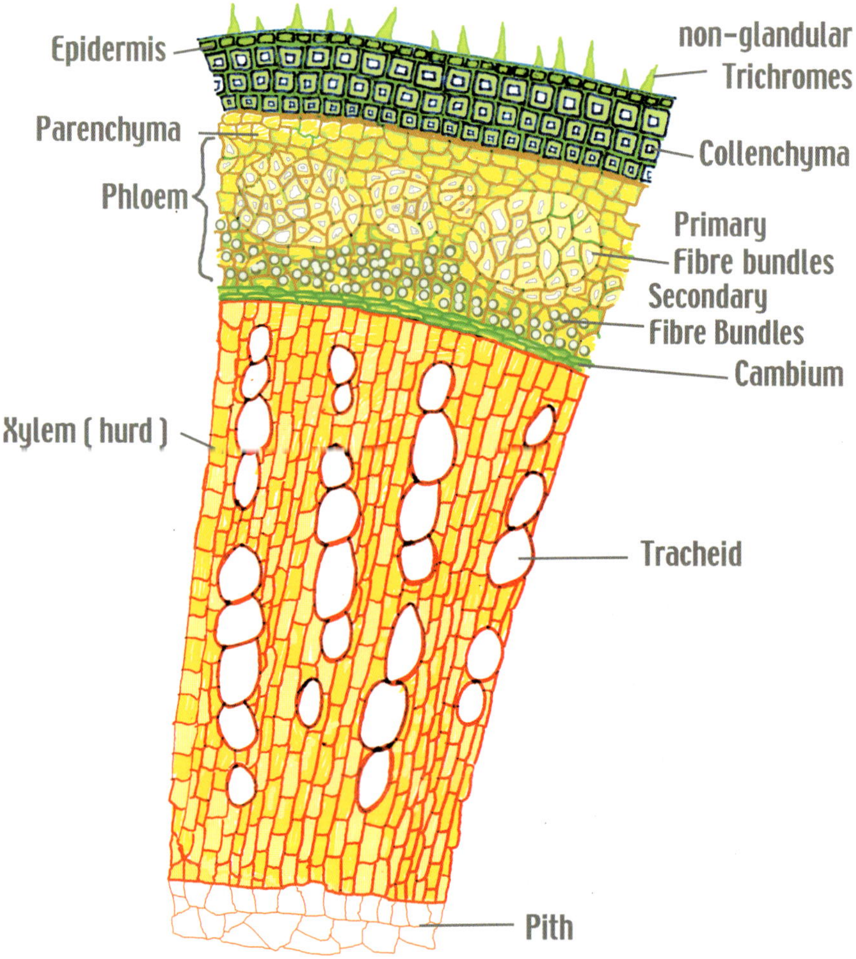

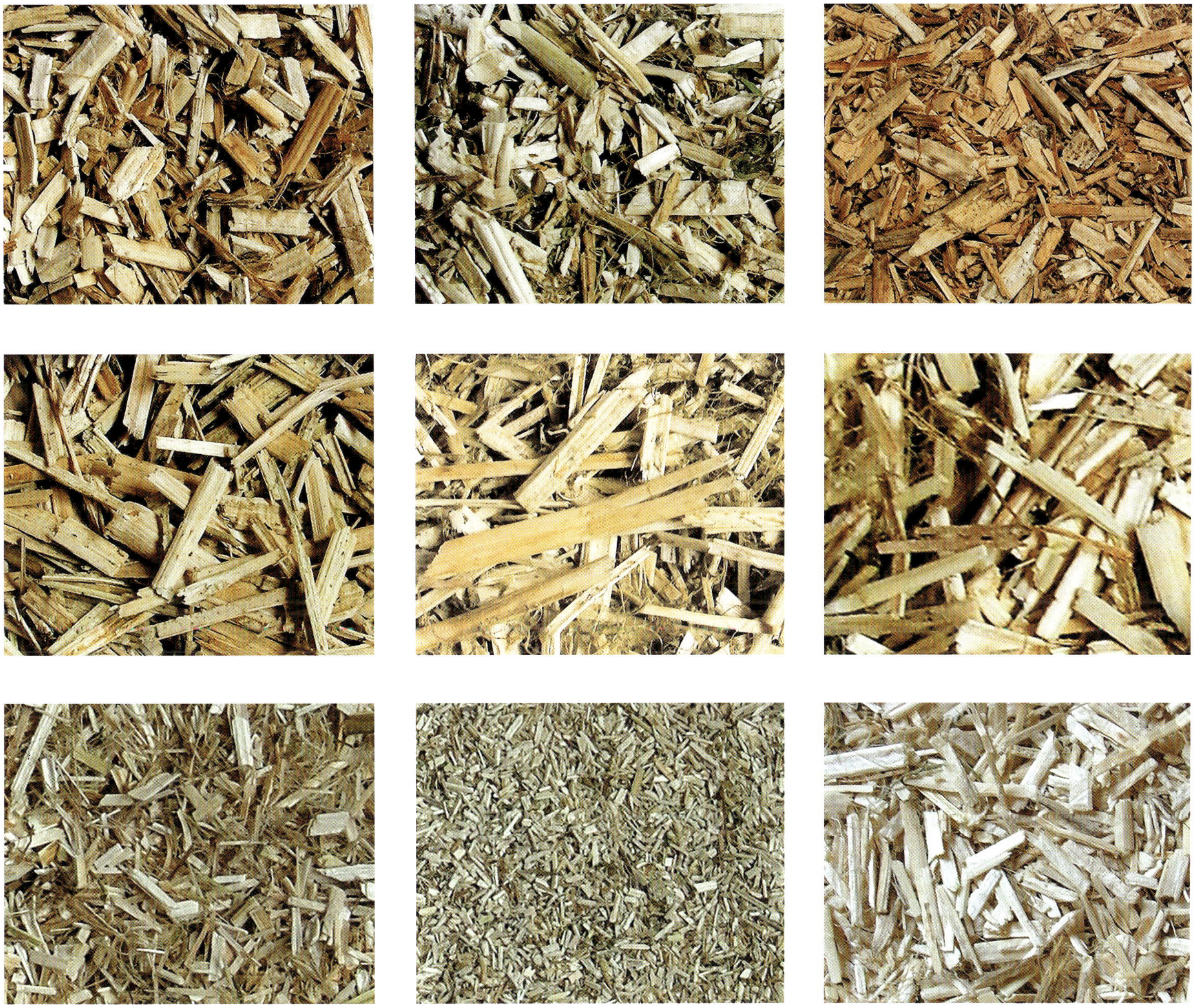

Variety of hemp hurd available for use as bedding and building material.

HEMPCRETE

Hempcrete History

In the Champagne region of France is the city of Troyes. As in many towns and cities in France, here there are numerous old oak frame buildings that are infilled with lime, straw and rubble.

This infill was normally covered with a lime render. However, in recent times these walls were repaired with disasterous results using cement renders. This meant the walls could no longer breathe and as a result moisture built up, causing the infill to swell and crumble and the render to pop off.

No-one I have talked or written to can remember the name of the mason who first used hemp as an ingredient in the mix that he used to repair these buildings with, but as Troyes is in the major hemp producing region of France, it is easy to see how he came across the idea.

At the time hemp was undergoing a revival of interest, and three people in particular saw the potential of this concept and developed it further. They were; France Périer (Isochanvre), Bernard Boyeux (Association Construire en Chanvre),and Yves Khün (Association d' Adam).

Since that time the idea has spread, especially in France, where today there are many hundreds of projects that have utilised hemp and lime to build new houses or repair old ones.

Opposite; Oak frame buildings in Troyes, France.
Below; Damage caused by cement rendering.
Below left; Close up of damaged infill.

Early Experiments

Whatever the precise details of the very beginnings of the hemp building system, without doubt the first people to successfully develop and promote the idea were France Périer and Francis Aujames.

Their imagination helped shape the basic ways in which hemp and lime might be incorporated in a variety of building systems and form a new system on its own, with its trade name of Isochanvre. An early example of a 'hemp house' was built near Tours, in France, and the product won many prestigious awards in the late 80s and early 90s.

However, their claims of having patented an almost magic process to treat the hemp in some way, to somehow mineralise the hemp hurds, before they were mixed with lime, were doubted as being necessary. Although their loose fill insulation product using hurds treated with silicates of sodium and potassium, was used with some success in a housing estate in Brittany. They were also unable to validate their claims of archeological evidence of bast fibres mixed with lime having been found in the structure of a Merovingian bridge when I contacted them some years ago. They have now ceased trading.

Subsequently many other groups of professional and amateur builders in France were trying their hand at using hemp in renovation or ecological building projects, with many versions of mixes being used.

Naturally this led to certain problems being discovered with different approaches, and has brought about the formation of associations to evaluate the experience of as many people as possible, so as to be able to promote concepts such as 'best practice' for hemp building as a system of ecological building.

Above left; Demonstration of repair to old building using hempcrete.
Left; An early example of a house built with hemp.

La Maison d'Adam

In 1995 the restoration of a prominent building in the town of Angers, situated in the north of the region of Pays de la Loire, was undertaken.

This building, La Maison d'Adam, was recently dated to the late 15th century with the help of dendrenographic examination of the timber used in its construction. It is an outstanding example of oak frame building of this era, standing on the place de Ste Croix at the top of the town a full six stories high. The frame itself is covered in carved figures in all sort of poses, some in love, some playing musical instruments, others in quite obscene gestures exposing themselves to the towns population.

The renovation project was evaluated by the CSTB under their ATEx programme together with other heritage bodies, and the cement based infills that had been applied since the 1950s to repair the ancient daub, were removed and replaced with hempcrete supplied by "Isochanvre". The building is now back in use and returned to its former glory.

Below; Views of La Maison d'Adam

Lime

As lime is an essential element in the ingredients of hempcrete, I feel it is worthwhile explaining exactly what this forgotten material is and how it works.

This sedimentary rock has been used as a building material for thousands of years. From the fine plaster surfaces of the tombs of Pharaohs or the frescos of Rome, to the mortar of viaducts still standing after two thousand years, it has served as a material for cementing things together or covering them up with decoration.

Since the late 18th century lime use in construction has declined, with the requirement to make stronger bridges for increased volumes of traffic, to the need for artillery proof defences for the military mayhem of the World Wars, when speed and strength were the important factors required of a building material.

The majority of lime produced presently is used in the manufacture of steel and glass and as an ingredient in asphalt, foodstuffs, toothpaste and for purifying water and waste.

Today the qualities of building lime are beginning to be appreiciated again as the energy used to manufacture cement becomes more expensive and the use of more sustainable, natural materials, which need to breathe, make a comeback.

It is in this context that we are looking to lime as an ingredient to mix with hemp hurds to make 'hempcrete". It has the flexiblity and porosity we need to function correctly and the environmental credentials to be described as a comparitively ecological material.

In fact, as you will see from the section covering the geology in this book, lime could be considered an organic material in that it is made from a material that was once living.

We now need to rediscover this very useful material, from the way its manufacture is adapted to the varying types of limestone found around the world, to the methods of its burning in order to create the best material for the conditions of use.

Opposite;
Background; In the beginning were shells.
Inset; The exposed limestone landscape of the Burren, Co Clare, Ireland.

Lime Geology

Limestone constitutes 4% of the earths crust, but how did it get there? For the beginning of the story of limestone we have to go back to the shellfish that created it.

These creatures from the minute phytoplankton, right up the food chain to the giant nautulus, that inhabited the seas long ago were busy growing their shells by combining an unstable metal called calcium, with carbon dissolved in the sea from the carbon dioxide in the atmosphere above. Limestone is created when particles of shells or coral deposited on the sea bed for hundreds or thousands of millenia, build up in layers, and with a combination of the weight of the shell material itself and of the sea above it, are forced together to form calcium carbonate rock.

As actor Tim Robbins' character in the film "The Shawshank Redemption" says, it is then a matter of "pressure and time". The deeper the sea, the more pressure will be exerted on this rock, and the more time it takes, the harder it will become. These sedimentary rocks may then be overlayed with other rock formations and be squeezed or bent into denser forms of limestones.

Most of these limestone rocks were made under deep seas where little other matter was deposited with the shells, creating a pure calcium carbonate rock. Where the deposits were occuring near the estuary of a river however, silt material such as clay was mixed in with the shells. This material will contain silica and aluminium, as most clays do. It is these impurities that will produce hydraulic limes, which when burnt will set even underwater.

Top; Coral sand
Middle; Diagram of limestone creation
Bottom; Limestone rocks: Left, chalk. Middle, Jura stone. Right, Hard mountain limestone

Quarrying

The act of quarrying rocks has become a highly contentious issue in many places. As the geology of an area contributes significantly to the appearence of the geography of a region, the removal of part of a hill-side or landscape is seen as nothing but destruction by some.

The process of digging down through layers of rock can have detrimental effects to not only the visual ammenity of an area, but damage water supplies and create a problem of dust and noise as well.

However, it seems to be an intrinsic part of human behaviour, from the 'feng shui' landscaping of Chinese emperors, to the desertification of nomads' goatherds, that we have been altering our environment to our perceived needs at any given time. If we hadn't we would all still be forest dwellers, and although some would say that that would be a preferable situation, it is nevertheless not the path we have taken.

At this stage in our species history we obviously need to take a more thoughtful route to our goals, if for no other reason than that population numbers are continuing to increase on a planet that does not. A reduction in any of the destructive ways in which we obtain the materials for living, never mind the energy consummed in the process, has to be considered.

The quarrying of gravel for use in concrete could be said to contribute more to the volume of material extracted from the earth for this end use, than the cement, and so to replace this material with one that is grown by farmers is already reducing the amount of quarrying needed to produce ones home. If a comparison is made with brick or concrete in the volume of a wall, then a hempcrete wall will only contain 25% of quarried materials in the form of limes used in the mixture.

Left; Limestone quarry, Co. Kilkenny, Ireland.

Lime Burning

Many technological developments have been made to the original methods of burning limestone rock in crude kilns to produce lime suitable for building projects.

Starting with small pits heated by charcoal, coal, or peat from beneath, to the revolving kilns of today, various changes to the fuels, the method of adding and removing the materials, have been made to ensure a more uniform quality of product and to reduce the costs of energy and labour consumed. The process is still on a chemical level the same, in that the limestone used is primarily calcium carbonate, which when burnt has the carbon in the form of carbon dioxide driven out. This leaves behind calcium oxide, a highly reactive and dangerous material in that when water is added to it gives off a great amount of heat and this needs to be done under controlled conditions. Another name for this substance is quicklime, which used to be supplied in rock form to be slaked (the addition of water) to produce calcium hydrate on site.

This is now considered too hazardous by all but those lime traditionalists who wish to recreate an historically correct procedure and have the expertise to carry out the process safely. This process, however it is done, produces either lime putty if still containing water or a dry powdered hydrated lime.

The hydrated lime is applied as a plaster or mortar and then starts to reabsorb carbon from the carbon dioxide in the atmosphere to become once more calcium carbonate. This is referred to as the "Lime Cycle".

Top; Modern gas fired kiln
Left; Old limestone kiln on mountain farmstead, west of Ireland.
Below Diagram of 'lime cycle'.

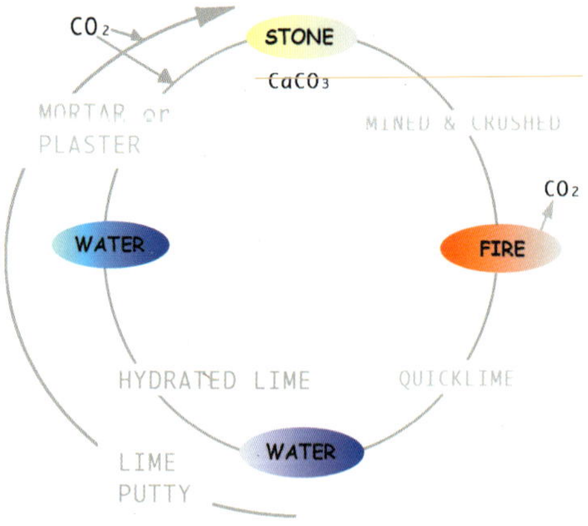

Lime Products

Most lime products packaged for the building trade are usually packaged as a dry powder in paper valve sacks, apart from lime putty which needs to be kept wet and away from the air.

Hydrated lime is available from most builders merchants, as it is commonly mixed in cement plasters to give plasticity when being applied and flexability when set. In countries where the use of lime in building is more common, hydrated lime, in powder form, might be available in a variety of qualities or premixed combinations as regards compression strength when set and speed of carbonisation.

Hydraulic lime in powdered form is only now beginning to be appreciated again in the construction world mainly as a material used in renovation, but recently in conjunction with other 'ecological systems' such as strawbale construction. Two major suppliers in Europe are in France 'Eurovente' and 'St Astier' but it is also made in Germany and Spain.

The manufacturers of hydraulic limes in France have been involved with the development of hemp building from the start and are now producing the first premixed lime products to be used with hemp.

Some are mixtures of cement and lime, some have more complex combinations of hydrated lime, hydraulic binders and pozzolans. The qualities needed of a binder to be used with hemp are that it should have a small enough particle size to give density and hardness to the lime element of the hempcrete and for the 'free limes' to be able to migrate into the structure of the hemp hurds on one hand, and on the other the ability to set throughout the material fast enough to be practical on the building site.

Added to this is the basic necessity for the material as a whole to breathe so as not to trap moisture within the structure and cause the hemp element to rot or for fungi to start growing.

Top; Lime putty
Below; Premixed limes used with hemp for casting and plaster.

Lime Mechanics

The structure of lime, both in rock form and as a lime putty is described as a colloid. This is a suspension of minute solids in a solution of even smaller particles as a continuous medium.
KEY:
Cream; calcium carbonate.
White; 'free lime'.

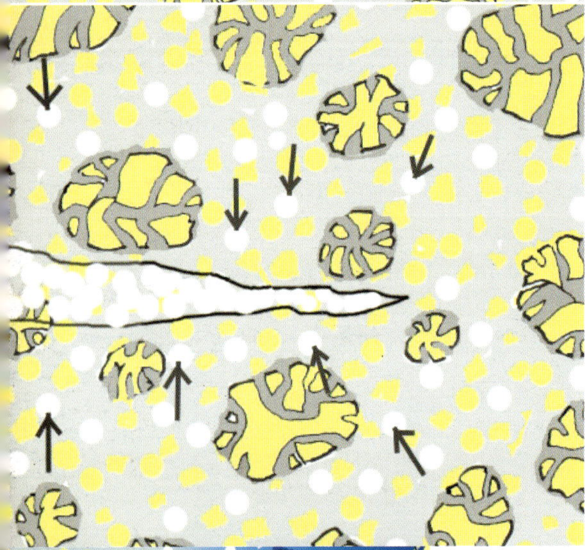

Amongst the carbonated fragments of lime are other tiny particles of calcium hydrate referred to as 'free limes'. These are transported about the mass with the movement of dispersing water vapour. If a small crack is formed in the lime based material then these 'free lime' particles will be deposited in the fissure and reseal it. This is called Autogenous healing.

Although being an extremely absorbent material, lime can also be used as a waterproof layer. When the surface is fully saturated only a small amount of additional water can penetrate, as a result of further absorbtion, behind the surface. The excess runs down the exterior.

Why is it so important to use lime instead of cement? How does it work? These are the questions one is asked when explaining hempcrete.

Recently magazines and television programmes devoted to self-build and renovation projects have highlighted the rediscovered qualities of lime.

I have always been attracted to the kind of descriptions of things known as "laymans terms" and so I will try to do this for you here.

To begin with we have lots of tiny little pieces of burnt crushed shell . As these are burnt the carbon dioxide makes small holes in these as it escapes, leaving behind spongelike particles. The porous structure of these particles is very important. These larger particles are floating in a sea of even smaller particles, whether they are a powder or paste, this is termed a colloid. As they then reabsorb the carbon they bond to each other to form a mass. However some of them are free and easy, and unmarried by carbon and are thus still moving about a bit through the material.

As the set takes hold a very open porous structure is formed which can hold a lot of water, but let go of it, just as easily.

However this uncarbonated 'free lime' migrating through the lime element of the hempcrete also passes in to the particles of hemp. These then coat the inner structure of hemp hurd cellulose and over time will replace it and turn the material into a fossil.

Cement

For most builders the suggestion that an old fashioned, slow acting material is superior to one that is modern and acts fast, is to be met with suspicion at the very least.

The fact that lime mortars have been used for thousands of years and that cement as we know it is only a fairly new commodity, make little impression on the present visitors to the local builders yards.

For the civil engineer, the invention of Portland cement as a superior hydraulic cement for use underwater and with a very high compression strength, has meant that impressive, and sometimes beautiful structures have been possible to build across rivers or towering over cities throughout the world, which could only have been dreamed of previously.

It has been in the field of house construction that cement has recently been attacked for its embodied energy figures and the quantity of carbon dioxide released during its manufacture.

Perhaps more relevant to hemp-building is its unappropriate behavior as a product to use in conjunction with other natural materials. It is the microscopic structure of Portland cement that is primarily to blame for this behaviour, and to fully understand the difference between the way in which cement and lime work we have to look at the manufacturing processes.

Both processes start with the burning of limestone (for cement the limestone is mixed with other minerals containing iron oxides) however with cement the temperature is much higher, reaching 1200-1280 celcius, compared to 900-1100 for lime products. This higher temperature is necessary to vitrify the lime to form nodules of clinker which are then ground to a fine powder.

The resulting materials setting action is then regulated by the addition of gypsum.

It is this vitrification procedure that creates the difference between the two substances. As we have seen on the previous page, the minute particles of lime are porous as a result of the carbon dioxide escaping, but the glass like cement particles are impervious and so do not absorb and then release moisture, but tend to be coated with a layer of condensed water. This might not be a problem if it were not for the fact that the particles are so small and densly packed together. This means that once the water has got in between the particles it has difficulty getting back out, or to put it another way, it cannot breathe.

The manner in which cement sets is of interest for creating hempcrete. However, as it is a chemical reaction resulting in a crystalline structure which happens without the need for carbon absortion and it is this that, if added to the lime mixture produces a complete setting action throughout the conglomerate.

It is getting the balance right in order to maintain the porosity that is important.

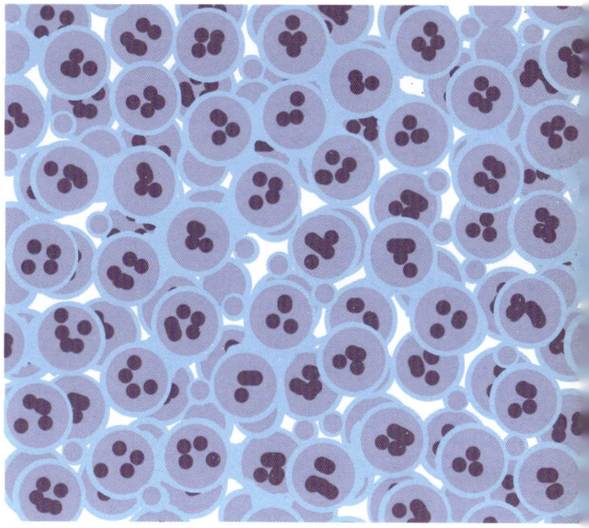

Above; Diagram showing cement particles coated with moisture.

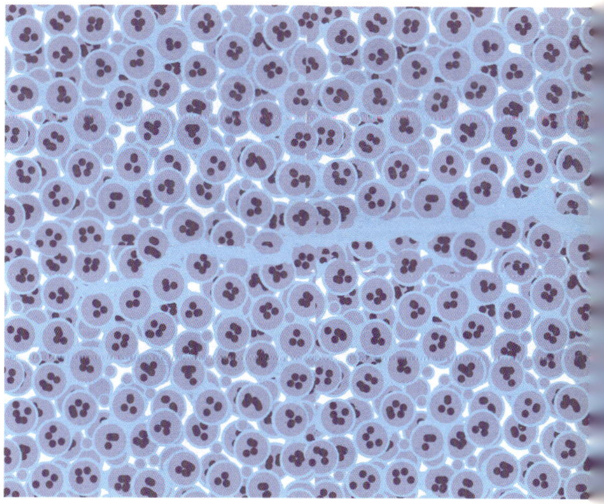

Above; Cement structure showing capillary action of water in fine cracks.

Types

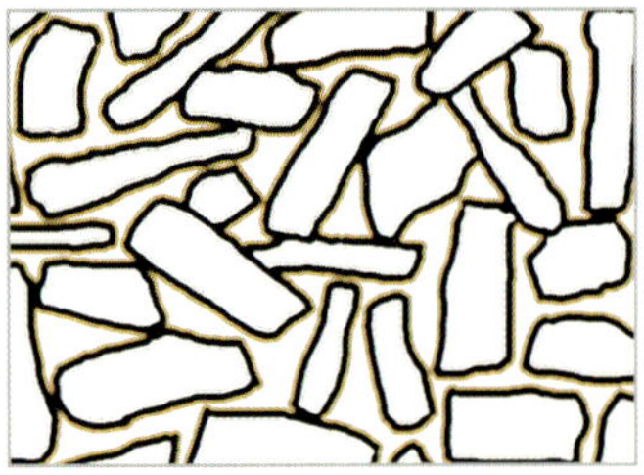

(i)

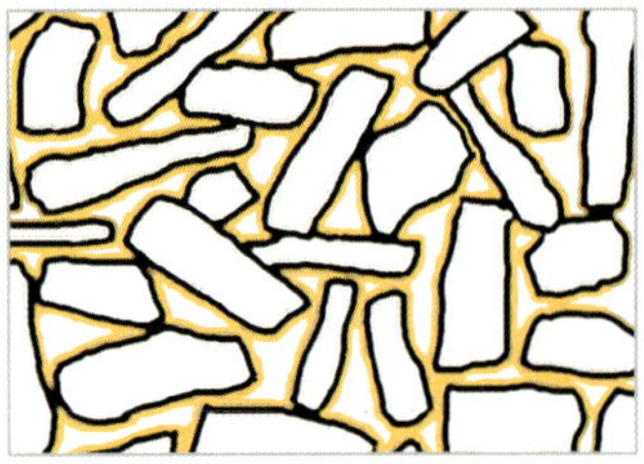

(ii)

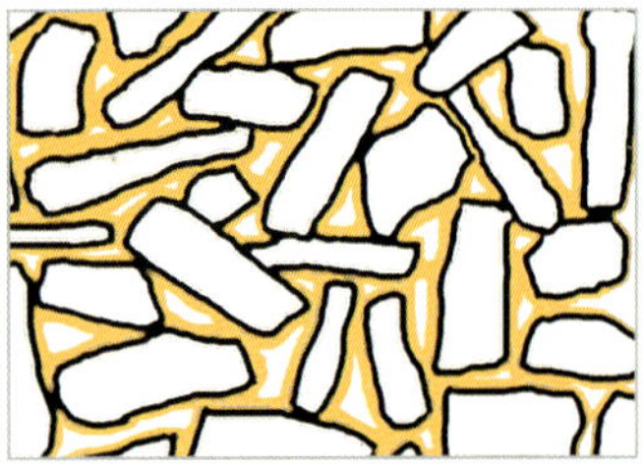

(iii)

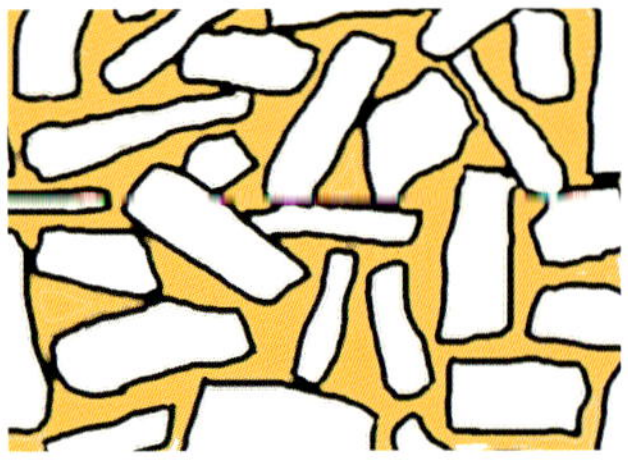

(iv)

Diagrams of lime coating in hempcretes.
(i) Lightweight hempcrete
(ii) Wall mix
(iii) Floor mix
(iv) Plaster

Through the trials and tests done by those experimenting with this hempcrete material, there are now four different mixes of hempcrete that have been evolved to be used in the construction of a house.

By combining the hemp hurds with varying amounts of lime, the elements of insulation and structure can be adjusted to suit the differing uses.

LIGHTWEIGHT HEMPCRETE

When used for its insulation qualities primarily, structure is of least importance as the hempcrete will be applied to areas where the material is only required to stay in place, such as loft spaces and between floors. This mixture has only sufficient lime added to the hurd to approximately 10% of the volume in order to coat the hemp particles and fix them to each other, so that the whole becomes a weak mass.

WALL HEMPCRETE

To be able to withstand the forces of impact, wind loading and the additional structural ridgitity likely to be needed by a wall, the amount of lime added to the hempcrete mix is increased to around 25% of the volume in order to more solidly bind the hemp particle together in a more rigid manner, but to retain as much of the insulation qualities as possible.

This and the previous lightweight hempcrete may contain up to 10% of the volume of hurd as fibres to add an element of reinforcement to the structure.

FLOOR HEMPCRETE

When used as an insulating material beneath a solid floor at ground level when slightly more compression strength is needed for the hempcrete to withstand the loading between the gravel beneath and the mortar and tiles above. Even more so when there is a screed to provide thermal mass for underfloor heating systems. For this use the lime portion of the mix is increased to 50% of the volume.

HEMP PLASTER

In order for the mix of hemp and lime to meet the requirements of a plaster, sufficient lime needs to be added to produce a workable material, and to bind the hemp hurds together to form a practicle surface. To do this, the lime content is increased to 85-90%. In some situations it is advisable to incorporate ground limestone into the mixture in order to increase the water retention abilities of the plaster in situations where it might dry out too quickly. Sand might also be added to produce a harder wearing surface or to enable it to be applied with a spray gun.

A smooth surface can also be obtained by using the sieved smaller particles of hurd to mix with the lime. These smaller particles produce a more plastic material which can be more easily smoothed with a trowel or float.

The four versions of hempcrete can be utilised throughout a building provided they are used in the correct way

The lightweight mix is suitable only as insulation, as it has no structural capabilities. It does however, provide a breathing insulation that unlike many other forms of fibre insulation does not sag or compress over time, it will also provide an element of fire-retardant. It can be used in roof spaces, interior partition walls, behind or between facings of external walls and between floors as a thermal and acoustic insulation.

The wall mix is strong enough to give added rigidity to the timber framework, but is classed as non-load bearing masonery in that floor joists or roof timbers cannot be supported by it. It is strong enough to help fairly minimal timber frames withstand considerable windloading.

As an insulation layer beneath solid floors at ground level, the floor mix is used. This contains just that bit more lime to prevent compression of the material under the loads per sq metre such floors would be subject to. To maintain breathablity in this position the flooring hempcrete is laid on a bed of gravel which needs to be slightly ventilated.

When mixed with sufficient lime to become a plaster almost all the spaces between the hurds are filled with the lime mixture, thus producing a more solid material that nonetheless still breathes and insulates. Provided it is applied to a wall constructed of a material such as brick, stone, or indeed hempcrete it will work perfectly.

Situations of use

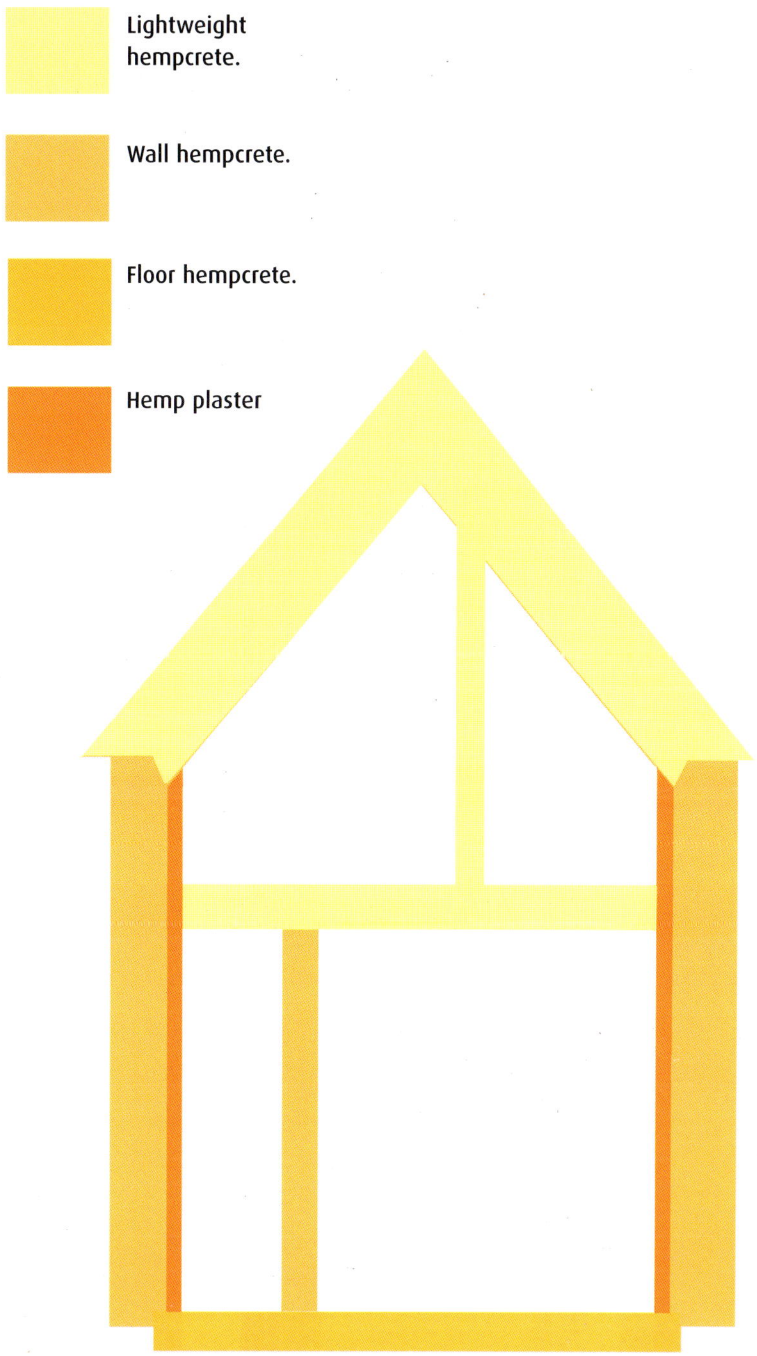

Gypsum

Gypsum is included here as it is used by some hemp builders, together with hydraulic lime, as an ingredient in the hempcrete mix to produce a chemical set throughout the mix prior to the carbonation process.

Promoted by Yves Khün, who operates with his architect wife from Avignon in the South of France. It is a method also employed in areas of Switzerland.

Gypsum occurs naturally as the deposits of ancient dried up lakes, also called plaster of Paris, as large deposits of high quality white gypsum are found in the surrounding area. Most people who have ever had the misfortune to break a limb will be familiar with it as the cast applied to the injury.

There is now a version of gypsum on the market often used in plasterboard which is the product of scrubbing or removing the sulphur dioxide from powerstation flue gases. It is produced when water is sprayed through the exhaust gases and this water is then filtered through lime which then becomes calcium sulphate. This material could be seen to be an ecological alternative to real gypsum which is an increasingly depleted resource.

Used traditionally as an ingredient for finishcoat in ornate plasterwork, gypsum is employed these days as a skimcoat plaster for walls and ceilings and to manufacture plasterboard. Its possibilities are limited when used in hempcrete as it does not breathe in the same way as lime does and the 'set' disintegrates when wet.

This narrows the use, with hemp, to climates where dampness is not an issue and humidity levels are low all year round. It would not do well in areas of windblown rain and tends to turn black if not well ventilated.

Top; Skimcoat
Bottom; Pierrefleur "Terrain d'Aventure", Lausanne, Switzerland, made of hempcrete with gypsum.

Hemp & Earth

From the earliest of human attempts at building, methods have been developed for moulding the walls of a building using the most basic of the materials to hand.

Earth and clay have been used in a variety of ways, both in the rammed earth system, using forms and shuttering and the more basic one 'cob'.

The term 'cob' is used to describe the building material made of a mixture of clays, gravels and straw, which is built up in lifts and then compressed by treading it down and trimming it off with a special spade/fork as you go. When moulded as a wall it gives a building thermal mass and insulation and is used in many diverse climates. In recent years many of these houses have been damaged in the same way as the old medieval oak frame buildings were when rendered or repaired with cement. Often whole sections of wall burst out from the surface as the moisture, locked in the wall by the cement, increases.

This type of building either has to be repaired using the same mixture of materials as were used originally, or alternatively by utilising hemp.

The hemp-lime combination can be used to both fill in holes in the wall or as a plastercoat internally. Cavities or indentations in the surface of the wall can be made suitable for repairing by wetting them with a limewash solution and then packing with a mixture of hemp hurds and clay or a mix of hempcrete normally used for floors. When plastering with hemp on a earth wall house, the friable surface of cob is made more suitable by the protruding strands of straw, while the more solid surface of a rammed earth wall can be used to spray hempcrete or hemp plaster onto.

In situations where the depth of material containing earth and hemp hurds might cause problems drying out, pre-cast bricks of clay and hemp can be used. They could also be used for internal walls or other structures not exposed to high rainfall.

Top right; Cob wall plastered with hemp plaster.
Right; Cob wall being built.
Left; Damage done by cement render to earth wall.

SCIENCE

Above; Houses in Montholier, France built for comparative testing.

Research

It is easy to make claims about something you have witnessed or would expect, however these things have to be tested, analysed and proved, to be accepted by most people, especially those in the worlds of finance and insurance.

To this end most countries have testing centres either run by government departments or as private enterprises. In France there are several such bodies.

Two of these especially have been employed in the role of evaluating the hemp building system and materials.

The C.S.T.B, Centre Scientifique et technique du Batiment (Center for building Science and Techniques) has run tests on the hemp-lime system and materials since their conception, especially in the range of insulation values and fire tests. They also evaluate materials and their suitability for heritage projects.

The other main body in France that has extensively tested hempcrete as a building material is the Ecole Nationale des Travaux Public de l'Etat (National School of Public works of State) and their Department of Building and Civil Engineering. Initiated by the French hemp building association Construire en Chanvre and under the direction of teacher and researcher Laurent Arnaud. Many comparative tests have been done between hempcrete and other materials for thermal properties of conductivity, capacity, effusity and inertia, also for compression and fire-resistance some of which are summerised on the proceeding pages.

Construire en Chanvre have also been instrumental in establishing training programmes for utilising hempcrete in construction at the Institute Universitaire des Métiers et du Patrimoine (University of Heritage Trades) in Troyes.

On the Industrial front Lhoist Construction have been researching the lime combinations best suited to use with hempcrete. More recently St. Astier have also come on to the market with premixed binders to be used in combination with hemp.

Above; Laurent Arnaud of ENTPE, and Michel Rizza of Lhoist at CnC conference, Châlons-en-Champagne, France.
Left; Maite Waag of IUMP, Troyes, France.

Hemp Houses at Haverhill

In 2001 a housing estate was built for the Suffolk Housing Society at Haverhill, Suffolk in the U.K.

The group of 18 houses, included 2 built using hemp, were from the outset to be used as a comparative test of contemporary building methods and materials, with the hemp-lime system.

Designed by architect Ralph Carpenter who had used the Isochanvre material and system for an extension to his own house, and monitered by the British Building Research Establishment Ltd a leading body in this field in the U.K.

Initially plagued by weather problems and further hampered by the labour force getting to grips with a way of building totally new to them, and with different materials, the houses were nevertheless completed satisfactorily and are now occupied.

Considering that the walls of the houses built with hemp are less than the width necessary to comply with building regulations at present, they perform fairly well when compared to the brick built houses next door.

Since these houses have been built, the proportions and ingredients of the hempcrete infill have been tested and improved on in other parts of Europe, especially France, and in many cases would be built with walls of thicker dimensions. This together with the addition of hydrated lime would, I feel, give the houses even better performance as regards heat loss and consumption.

However, the main improvements that could be made in the technique would have to be in the training of the workforce and in the organisation and methods employed in the construction. It is definitely not economic to have two teams of workers; one only prepared to move the shuttering and another whose tasks are limited to installing the hempcrete.

The project generated a lot of interest and was reported widely in the press at the time.

A full report on the project is available on the internet at www.bre.co.uk and further information at www.suffolkhousing.org.

Opposite; Hemp houses at Haverhill, Suffolk, England.

Haverhill Tests

Of all the aspects of the tests carried out on the hemphouses at Haverhill, the pictures of the exteriors of the houses, built with both types of construction systems and taken with thermographic cameras are the most remarkable.

Reproduced below they show clearly the weaknesses' of the conventional method, especially as the hemphouses weren't even built to standard, insulation-wise, as far as the wall width was concerned.

The main points of weakness in the hemphouse seem to be along the foundations and the top part of the 1st floor wall under the eaves.

These are simple to rectify at the design stage by specifying better insulation at these points.

Figure 1: Front elevation of the Masonry House at Haverhill, Suffolk.

Figure 2:
Thermographic image which represents the front elevation of the Masonry House as shown in Figure 1: Colour graduation represents temperature scale of -3°C to 7.3°C.

Figure 5: Front elevation of the Hemp House at Haverhill, Suffolk.

Figure 6:
Thermographic image which represents the front elevation of the Hemp House as shown in Figure 5: Colour graduation represents temperature scale of -3°C to 7.3°C.

Thermal values

Many people are now aware of the rating system for thermal conductivity as the "U" Value or in America the "R" value given for a particular component or combinations of materials to be used in the house they are building. However, not all people are exactly sure what it means.

As the term implies, it refers to the ability of a material to transmit heat by conduction. This is calculated by applying heat to one side of a cubic metre of the given material and measuring the time it takes for the temperature of the opposite side to increase by 1 degree. This particular aspect of a materials behaviour is especially important for the manufacturers of toxic insulation materials, as it is possible to achieve very good results for these types of materials.

However, as with all such tests they are done in a laboratory, under controlled conditions and by scientists. Of course no normal house is built under such conditions and factors such as design, geography, climate and the correct application by the workers on site will all have an effect on the final result. Although the conductivity of a material is important in conserving the energy used to heat the building, there are health implications to living in an insulated box where the only heat is in the air, especially connected to respiratory ailments. This is often the case in houses built with materials that have good insulation values but no thermal mass (heat stored in the structure).

Hempcrete nevertheless has a very respectable conductivity rating or "U" value when used at recommended thicknesses with the added advantage of being safe to live around.

On the next 3 pages we will be describing the thermal behaviour of hempcrete with diagrams that show comparisons between different building materials.

The key to these materials is given below.

The background is shaded to represent the gradation from hot to cold. Orange = hot and Blue = cold

Concrete
Fibreglass
Aerated cement block
Clay brick
Dense clay brick
Softwood
Hardwood
Hempcrete (lightweight)
Hempcrete (wall)
Hempcrete (floor)
Hemp plaster

Thermic Capacity & Inertia

The qualities of thermic conductivity and capacity combined gives a material what is termed thermic inertia.

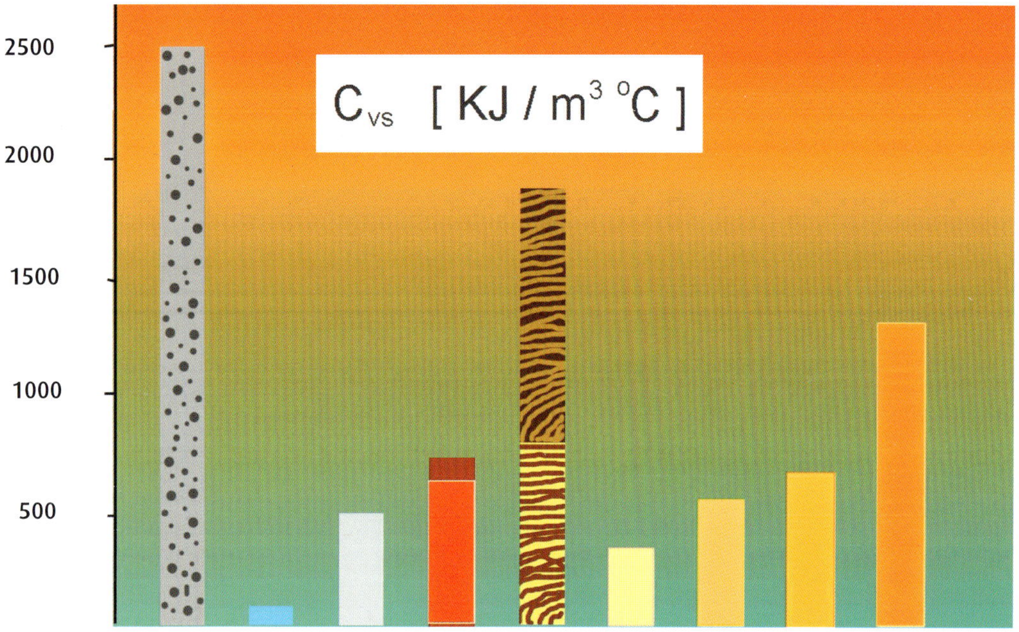

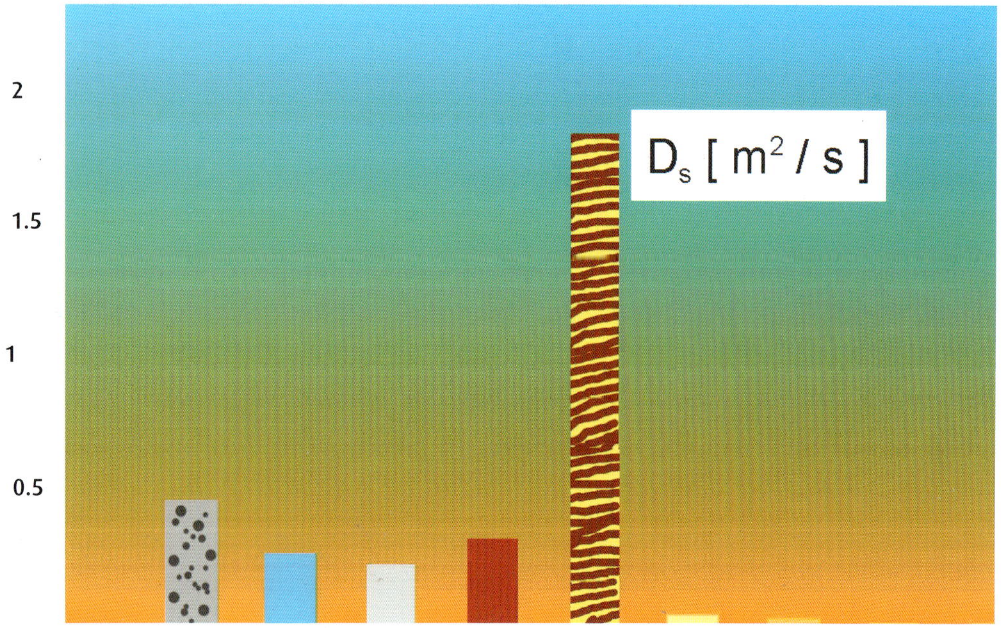

This is a quality that is now being better understood when analyising how a building reacts to the changing temperature outside, in relation to the temperature inside. As we can see from the diagrams, hempcrete gives a favourable result when compared to similar materials such as brick or aerated concrete blocks when it comes to capacity. Too high a value will result in a quantity of heat that is too great for comfort and difficult to control. From this we can see that the thermic inertia of a building is also affected by its mass. When the results are combined to give us the inertia value, hempcrete does very well in providing an even temperature and performs better than most other types of materials.

The quality of inertia works with heat generated either inside from a heating appliance, or outside the building from the suns rays. So this phenomenon is equally important in the summer as it is in the winter.

To calculate the thermic inertia of a material is fairly complicated. Basically, to have a good thermic inertia, a material must have a high thermic capacity and a low conductivity, two tendencies that are difficult to reconcile. Humidity will also have an effect on these values as will the latent heat of the moisture.

Effusity

Another aspect of a materials behaviour that needs to be included in order to understand how it will feel to live in a particular building is the thermic effusity of a material.

This relates to the ability of a particular substance to release the heat stored within it from the surface. Again, hemp does well when compared to other materials in the test done for CnC.

In the diagram on the right we can see that heat stored initially in the surface of the material readily flows back out again, once the air adjacent starts to cool. This happens even with the heat generated by the occupants own body heat.

This phenomenon greatly effects the 'comfort' one feels in proximity to the wall. The temperature of the walls is important in that the temperature felt is the average between the temperature of the air and that radiated by the walls. Neither the conductivity or the thermal capacity of the hempcrete reflect the sensation of comfort obtained in the nearness of a particular material.

To explain the calculation of the effusity of a material; Effs = the speed with which the superficial temperature of material evolves. The more elevated it is, the more the material will be considered subjectively cold, the value of the wall would be much lower than that of the body and would result in an uncomfortable exchange of heat by radiance. When it is low the radiance is weaker and the body is able to maintain thermic equilibrium more easily as the material is then felt as "subjectively hot".

Heat given off by person in room or other source is initially absorbed by the wall and then later released back into the room

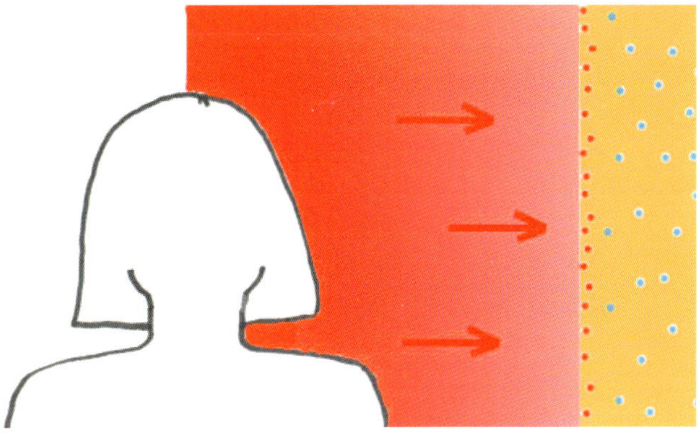

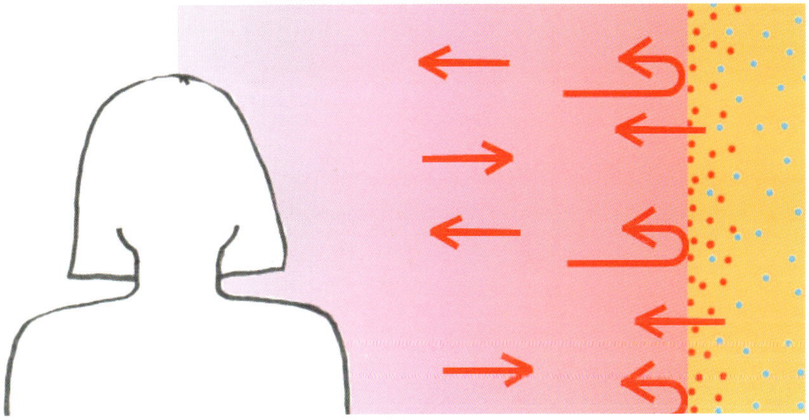

Water

Below; Test rig built by the BRE Ltd to test water-penetration of a hemp wall.

As the system of building with hemp utilises the solid wall system, most people's initial reaction is to ask does the wall remain waterproof without a cavity or waterproof membrane?

In order to satisfy this question the BRE in Britain carried out water-penetration tests to evaluate the ability of a hempcrete wall to resist rainwater penetration.

To do this they made up panels of hempcrete 50 cm x 50 cm and mounted them in frames around a water spray device. The surface of these hempcrete panels were coated with a lime render and limewash to replicate the finished surface of a hemp wall and were then subjected to simulated driving rainfall for various lengths of time.

During the tests carried out on what might be considered normal amounts of rainfall (ie 2-24 hrs) the penetration beyond the layer of lime render and into the hempcrete material was only of the order of 20-40 mm and for 96 hrs only 50-70 mm. However, this was with higher than normal quantites of water for most geographical areas, and during a 96 hr period the quantity of water represented up to five years continuous rainfall for some regions.

We can presume with any quantity of rainfall it will not be continuous and that the wall would dry out again in between the downpours.

The presence of water in the hempcrete in levels of above 20% moisture content for any period of time will be bound to create problems. As with any natural material used in the construction of a house, if damp problems persist, wet or dry rot will set in. With a hemp building this could happen to both the hempcrete and the timber frame.

Fire

The ability of a building material to resist fire is obviously very important. However, to inhabit a house built completely with materials that do not burn would be very uncomfortable and cold, as most natural materials will burn when subjected to sufficient heat.

What we can try to ensure is that if there is the disasterous event of a fire in ones home, that the fire will not spread quickly and that the smoke and fumes given off during the fire are not poisonous or extremely toxic.

As both the hemp and the timber frame are obviously flammable, the design and position of these components will be important in ensuring that the house or building is not unnecessarily dangerous.

The particles of hemp in the hempcrete material are all coated with a layer of lime which in itself is not combustable at normal house fire temperatures and so as a mass is rendered safe. Tests done on a variety of mixtures and methods of use of hemp-lime materials have shown hempcrete to be nonflammable (that is they smoulder or burn with no flame) while the material is in the drying out phase. When completely dried out it is classified as non-combustible.

The method of using hempcrete together with a timber frame usually entails the positioning of the framework in or surrounded by the hempcrete material and so is enveloped by a largely fireproof layer and is rendered safer than many structural timber elements would be.

Though no tests have been carried out with fire on a complete house built using hempcrete, there has been an unfortunate example of the resistance to fire of hemp plaster. As a result of a vicious arson attack on a nearly completed home, built using many natural materials and sustainable systems, the hemp plastered cob walls were the most undamaged of all.

DESIGN

Orientation

To design a building that is to be used as a home or workspace, it is essential to consider the sun. All to often we pass streets and groups of houses that have been positioned only with respect to maximising the number of buildings on a particular site, and thus the profit to be made.

To not take into account the elements of sun, wind and rain can seriously affect the amount of light and heat that will be needed to be added to comfortably and economically live in it.

The ecological aspects of design will therefore not only encompass the materials used in its construction, but how to best use those materials in the given environment.

It is worthwhile stating at this point that when one is using natural materials to build ones home they are of course biodegradable which means that if not used properly they will rot!!

When using hemp and lime it is advisable to assist the breathing aspect of the material by bringing heat into the building in sufficient amounts to maintain the correct humidity in the material.

Any moisture traversing the hempcrete from exterior to interior from rainfall, or vice versa through vapour produced by inhabiting the building, needs to be drawn or driven to the cooler surface and then removed by moving air.

This works in cooler climates by bring the heat from the sun into the interior and storing it for as long as possible in order to encourage the moisture to condense on the exterior. In warm climates the insulating aspect of hempcrete can keep the interior cool and maintain a certain level of humidity, however this will need to be controlled by ventilation.

Whichever of these situations applies to the project, positioning the building with regard to the sun will be an important factor.

Below; Diagram showing the ideal position of the building in relation to the suns' path

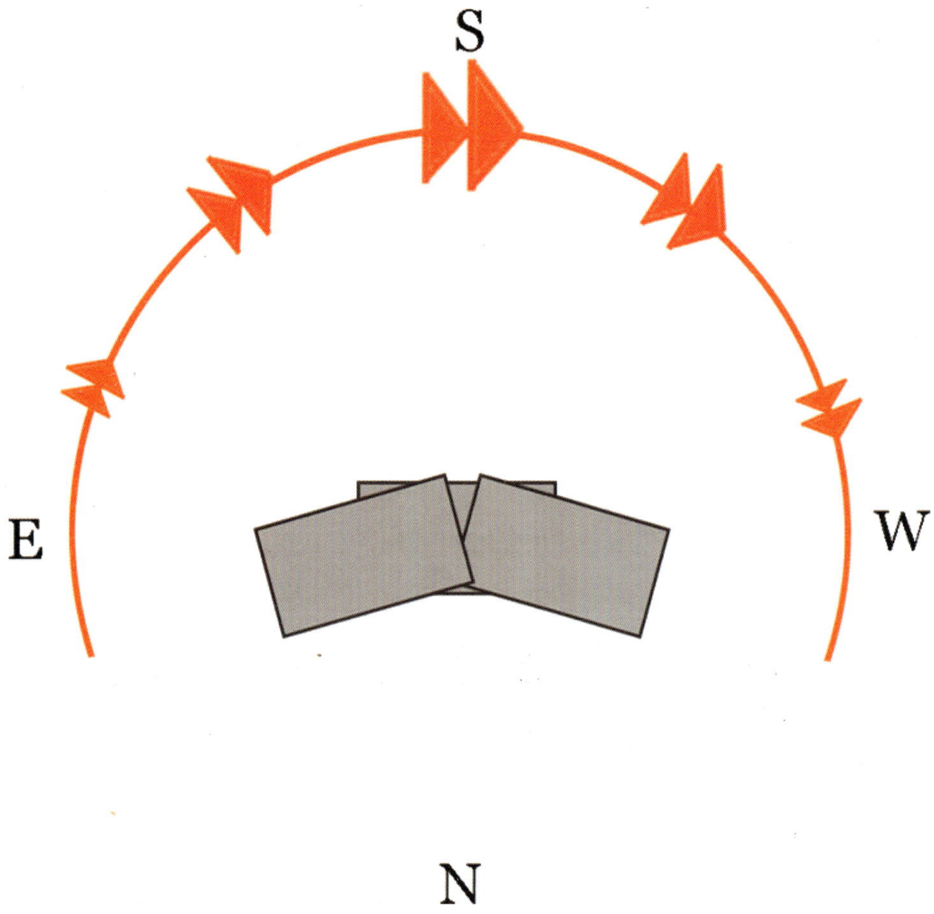

All the natural heat and light that can be utilised by the building should of course be considered at the design stage for both health and economic reasons. In many instances the traditional or vernacular designs of the particular locality are worth taking into account as they will have been evolved with these factors in mind.

Of course the past economic situation of a region might have limited the use of certain materials. Others, such as double glazed windows with high levels of insulation, would not have been available. So taking all these things into account, the aim is to incorporate the wisdom of the past with modern technology.

During the winter months the lower angle of the sun can be utilised by positioning windows facing towards the sun opposite a wall that can store the heat as thermal mass.

While through the summer period, light and heat will be radiated from the sun from a higher angle, and so will enter the building from above through any skylights there are in the roof and will be absorbed by the floor as it shines down.

The compatibility of materials with the hempcrete is of course important when they will be in contact. With timber this means that any likelihood of warping or splitting should be avoided and areas should be masked if the timber is liable to staining from the alkaline nature of lime.

Timber components should be used with regard to their breathability and so laminated strand boards should be used internally and more open structured timber used externally as cladding, or within the hempcrete where the lime can migrate into the structure.

Again with mineral products breathability is an issue as the terracotta tiles used on a ground floor will need to breathe, in order to allow the moisture below to migrate through them, whereas ceramic tiles will need to resist water when used around baths and sinks, but be stuck to the surface using breathable lime mortars and grouts to allow any moisture behind them to escape.

The sun shines from a lower angle in the winter and as a result windows should allow both the light and heat to enter the room and the heat to be stored in walls that it shines on.

As the summer sun both lights and heats a building from above, this energy can be utilised or controlled by the use of shading. Below; Light brought into the kitchen of a hemp house.

Hardwoods

Oak *(quercus)*

Of all the species of trees that produce timber used for constructing houses, none is stronger or more durable than oak. It is still possible today to see thousands of such buildings all over Europe built using oak as the framework.

With many varieties in the family, it grows throughout Europe and America. The acorns are readily fed upon by pigs and the bark contains high levels of tannin which has been used in the tanning of leather from ancient times.

Its timber is extremely hard but with an element of flexibility. Although slower growing than most timbers used in construction, it nevetheless makes up for this with its durability, lasting for hundreds of years if maintained properly by preventing the ingress of unwanted moisture.

Despite its hardness the wood is easily worked and was often used green to build frames for houses as the joints would tighten up as the timber dried i*n situ.*

This method could be used in conjunction with the hemp system, as the hempcrete will withstand a small degree of compression with no ill effects.

There are many other uses of oak in hemp building apart from the frame, as it is also suited to flooring and panelling if correctly sawn and seasoned.

Chestnut *(castanea sativa)*

Though this tree is probably better known for its nuts than its timber, it is still used for poles in agriculture supporting hops and grapevines.

It is very similar to oak, though it is generally not left to grow for as long, as the wood tends to split when the tree grows very large. For this reason it is normally coppiced and used as poles or split to be used as shingles.

Top; Oak logs in a managed forest.
Left; Oak frame houses still functional despite the effects of time, Troyes, France.

Softwoods

Douglas Fir *(pseudotsuga menziesii)*.

Though not actually classified as a fir, it is sometimes referred to as Oregon pine.

This widely grown tree is well known for its strong durable timber which is dense and fine grained. Due to the great size it can grow to, it is available in longer lengths than other timber might be. Ideal for use as uprights and beams in a building framework and especially suited to hemp building, due to its dimensional stability, which means that it can even be used green without the risk of too great an amount of shrinkage.

Scots Pine *(pinus sylvestris)*

This was the most common and widespread of trees in the northern hemisphere. The timber is sometimes called Red Deal and is primarily used internally as it is not known as being particularly durable in exposed conditions. However, it is perfectly suited to be used as internal framework as it is a strong fine grained timber with a good colour, especially the heartwood which is darker than the softer sapwood.

Larch *(larix)*

There are three species of larch; European, Japanese and a hybrid of these two. The hybrid is the most popular for timber production as it has inherited the resistance to larch canker from the Japanese variety, but is faster growing. Considered to be as good as oak in many respects but resistant to woodworm and much faster growing, larch is ideal for structural timbers or cladding in hemp building projects.

Sitka Spruce *(picea sitchesis)*

The most commonly planted tree for low value timber, sitka spruce is nevertheless perfectly suitable for studwork, if seasoned, to be encased in hempcrete as its open grain can absorb the lime.

HEMPHOUSE DESIGN; OAK FRAME

This style of construction has been around since medieval times and many buildings of this era are still standing throughout europe.

The oak framework would have been originally infilled with, either a wattle and daub combination, or a mixture of building debris of straw and lime. Later this might have been replaced with bricks.

More recently there have been modern versions made available that still utilise an oak framework, but are infilled with modern and usually toxic insulation materials. These new houses have the same appearence as older types and are popular with planning authorities in areas where the traditional styles of local architecture are being preserved.

The system of hemp building is ideally suited to this method of construction either when building a new house or repairing an old one. The framing system itself offers not only an attractive exterior, but has the added possibilities of the cantilevered frame and the feature of exposed timbers internally.

With a renovation project it will be necessary to cast the walls to a greater width than they were previously to provide sufficient insulation values and for external appearance, this is usually done on the interior of the wall creating a more modern look to the rooms.

Often the upper living areas in the building extend up into the roof area, and as it is difficult to raise the height of the roof structure, an insulation material that does not take up much of this space will need to be used. In the diagram I have shown hemp fibre insulation panels.

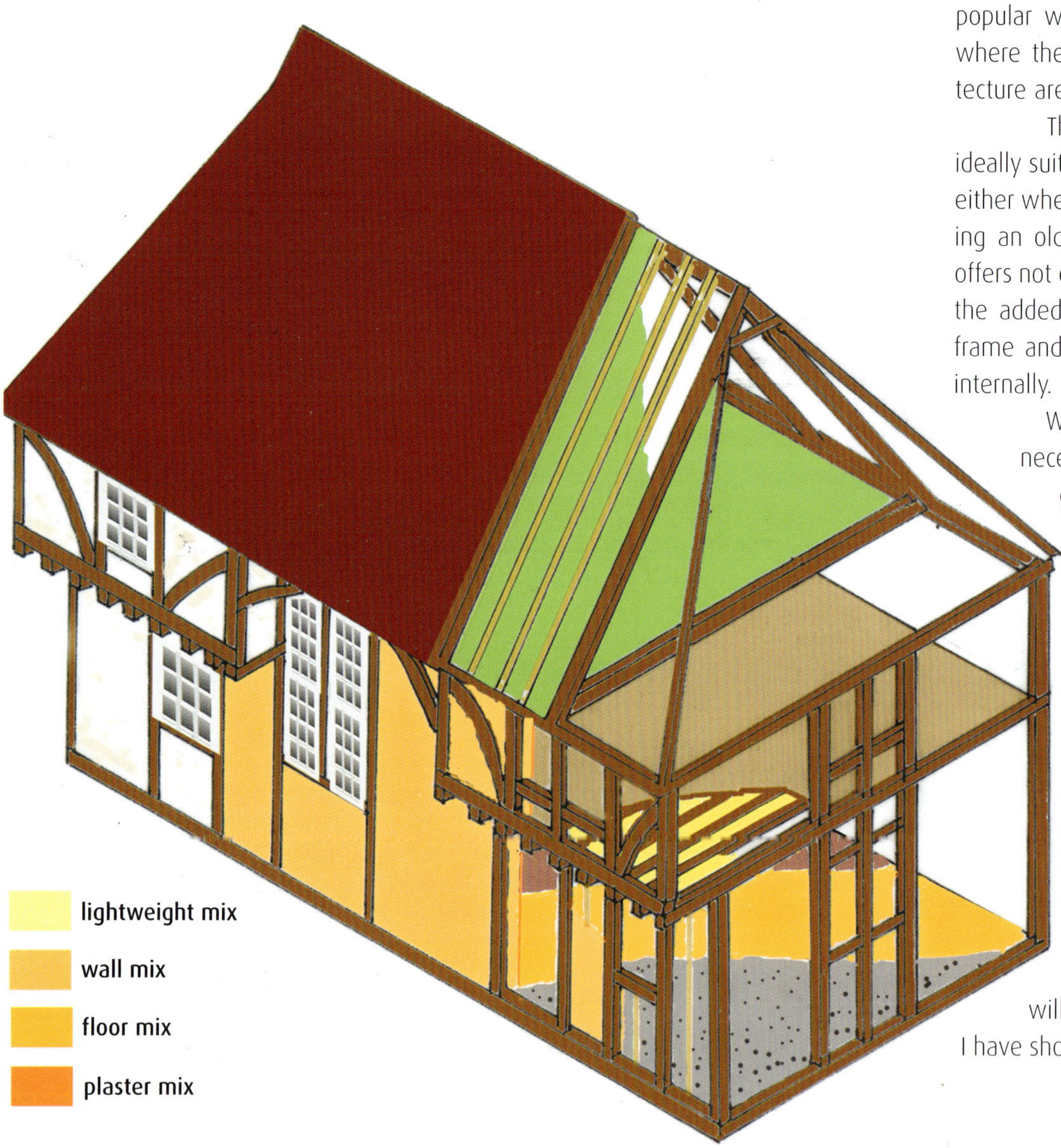

Left top; New light oak frame with double framework as feature on interior and exterior face, infilled with hempcrete.
Right top; Interior of Maison d'Adam
Right; Exterior of Maison d'Adam
Left; Oak framed building showing the use of cantilevering.

HEMPHOUSE DESIGN; POST & BEAM

With the post & beam type of framework larger timber elements can be used, thus reducing the energy used to cut the wood to smaller dimensions. This also reduces the number of pieces of timber used in the structure and of course the time to assemble them.

As the posts of the frame can be large sections of timber, they may often span the width of the wall and so form a surface to fix the shuttering onto, or provide a levelling point when spraying the hempcrete. When finished with an external protection such as a paint or preservative, the exposed timber framework may become a visual element of the exterior of the building.

This is also an ideal frame work for a building to be finished externally with heavy boards.

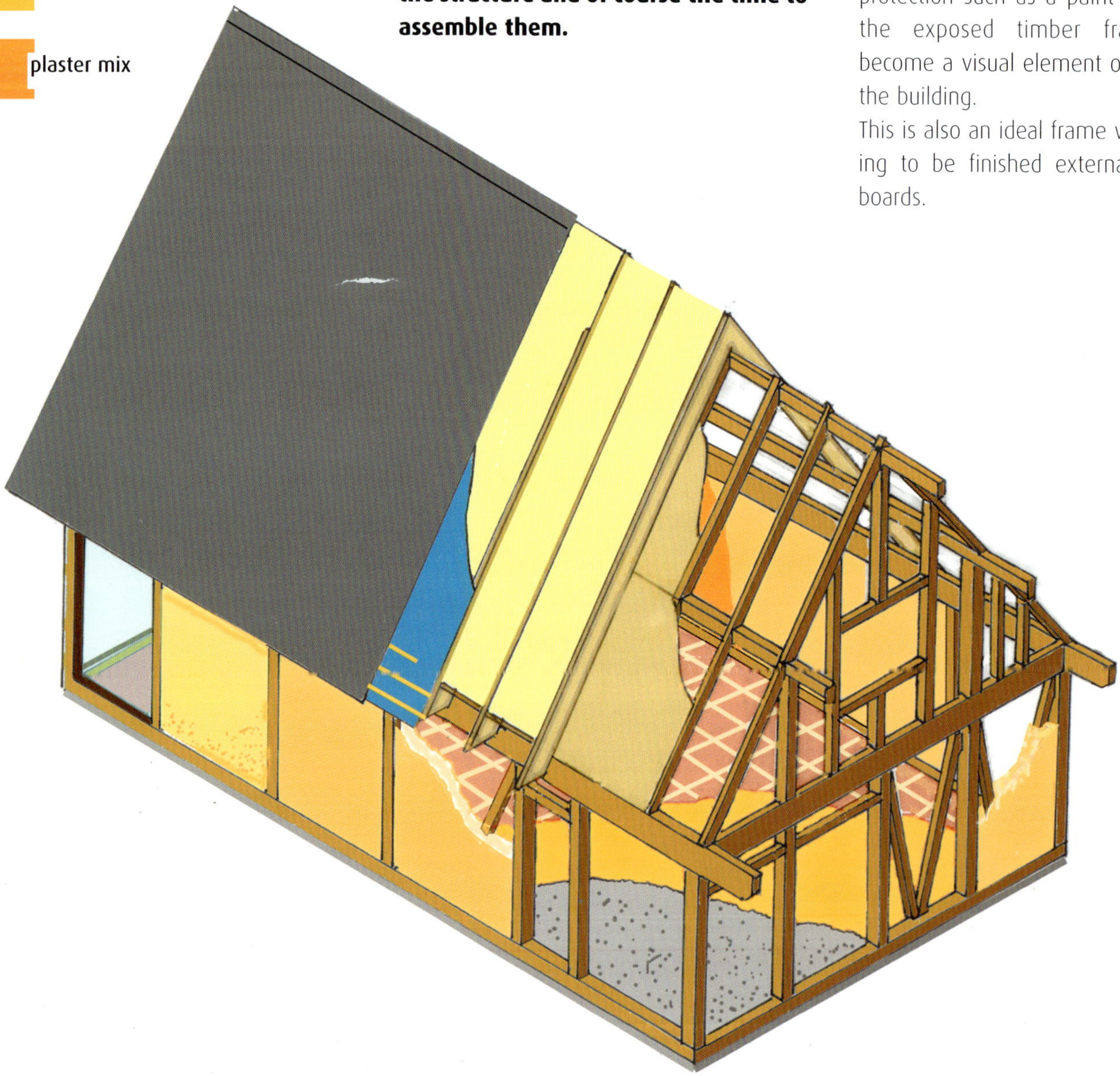

Right top; Soffit of rafters finished with timber panelling.
Right bottom; Timber frame with sprayed hempcrete infill.
Below; Frame being manually infilled with hempcrete.

HEMPHOUSE DESIGN; STUD FRAME

A stud or balloon frame is the most common style of framework used to build houses nowadays. It is a form of framework that equally lends itself to on-site erection, or as it is more commonly called 'stick build'.

Alternatively, it can be prefabricated and delivered to the site to be assembled *in situ*.

The dimensions of the timber used in such a framework will vary, usually from 100-150 cm x 50-75 cm. This allows them to be encased in the hempcrete wall in a position where the hempcrete completely surrounds them. They can also be placed where an inner surface or outer face of shuttering can be fixed to them.

This form of framework is also ideally suited to creating curved walls where these are desired. The upright elements are placed closer together as the radius of the curve decreases. This both gives strength to the structure, but also provides sufficient positions to fix and bend the shuttering boards to.

As all the dimensions of the frame elements are too small to form joints that would be able to take lateral force, it is necessary to include diagonal bracing members in the design of the framework in order to make the whole thing more rigid.

At the corners it is also advisable to double up the uprights to provide sufficient strength at this point.

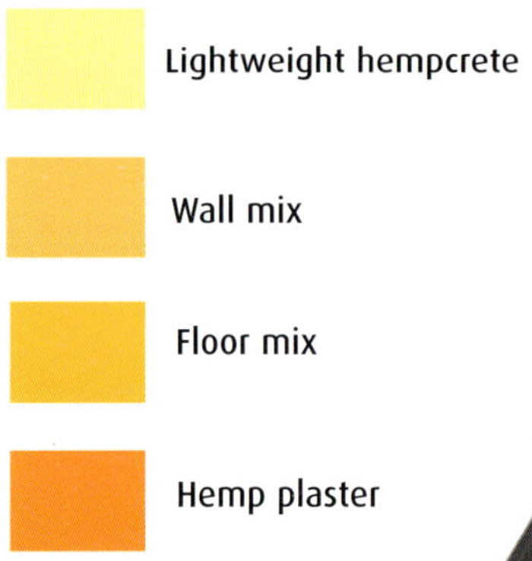

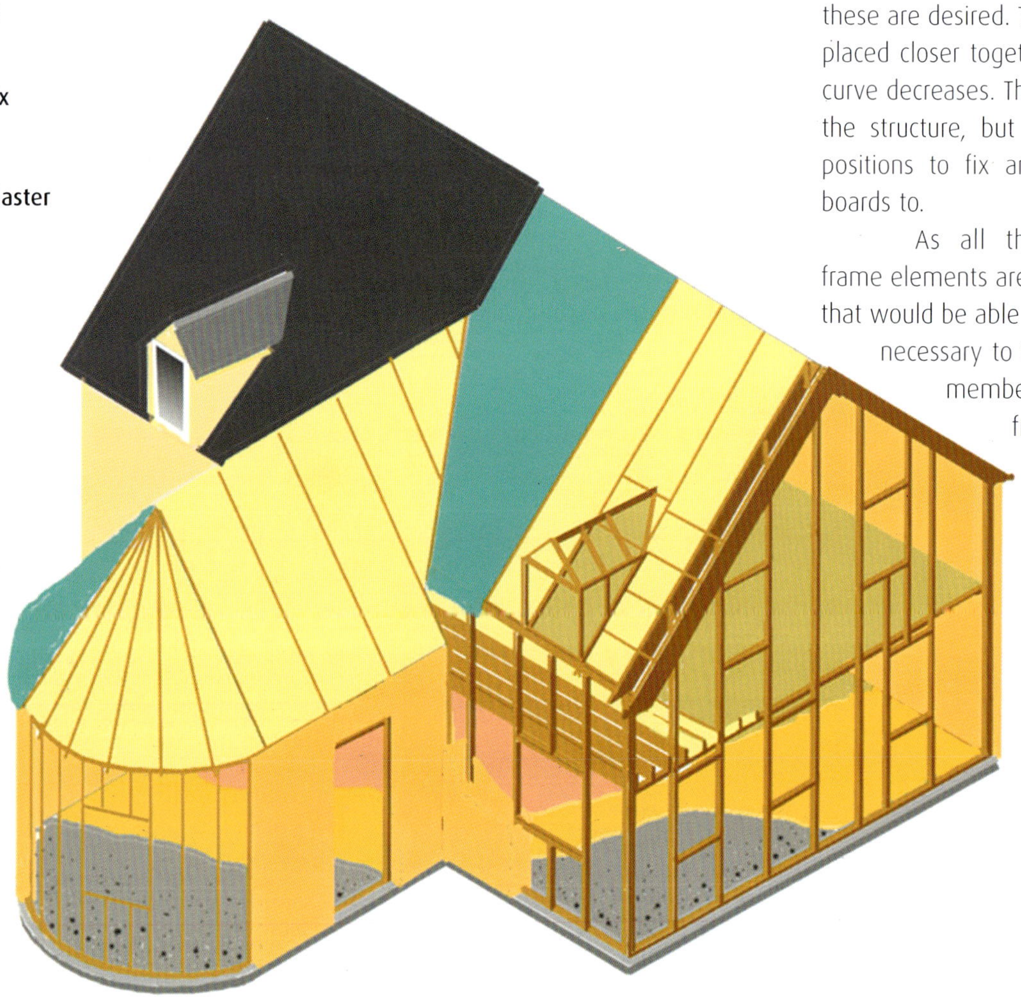

Above ; Stud frame with diagonal bracing.
Right; Hemp house with stud frame embedded in walls finished with lime render and timber cladding.

HEMP HOUSE DESIGN; LIGHTWEIGHT STUD WALL FRAME

This combination of frames has been developed for hemp building by Claude & Catherine Eichwald, architects who work in the Alsace region of France.

It incorporates a minimal use of wood with simple detailing around doors and windows.

In the centre of the structure is a strong post & beam type frame that both supports the roof and floor and is used to fix a lightweight exterior spruce frame of 100cm x 50cm elements to. This lightweight frame is faced with Oriented Strand Board to form the interior wall surface and removes the need for shuttering internally.

To the outside of this frame work is fixed heavier frames at the apertures of doorways and window spaces. These are of sufficient size to finish at the outer level of

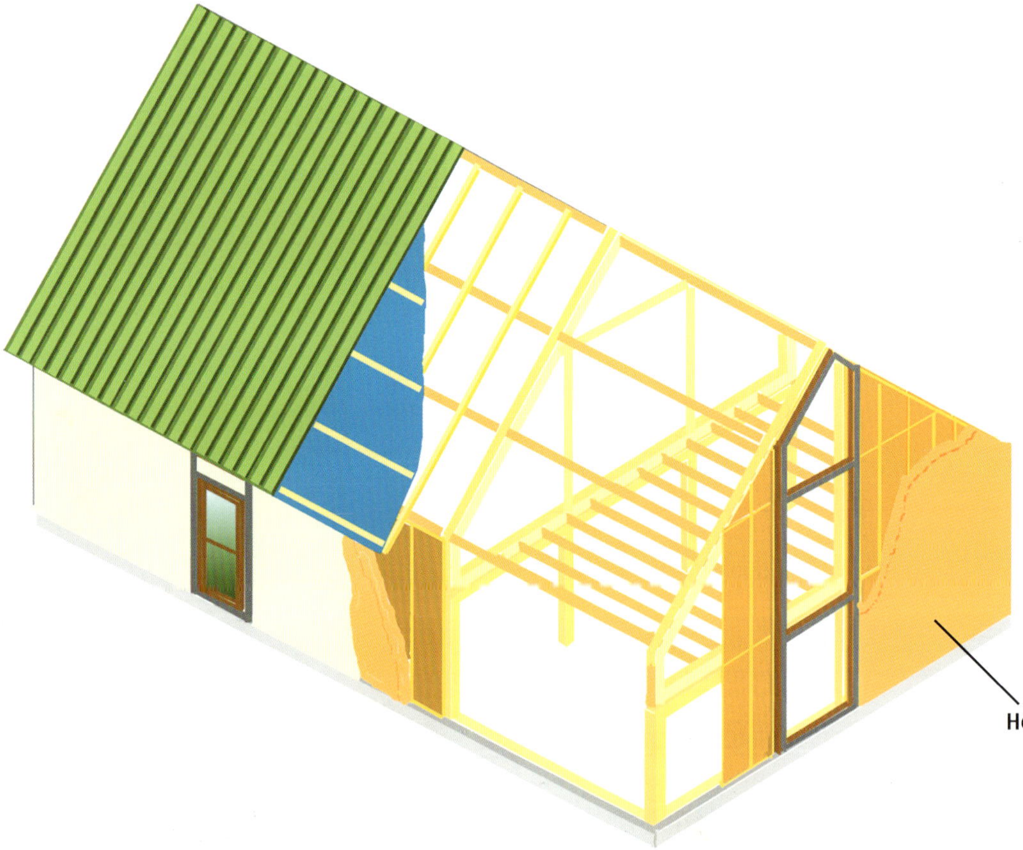

the exterior hempcrete wall and to be rebated to take the window and door frames. This exterior frame should be made of a timber such as larch or other suitable wood for external use.

The internal supporting framework could be made of round poles or salvaged timber to further reduce materials, and the energy used to prepare them and to create a feature to the appearance internally.

Right; Round poles used as internal supporting frame.
Below; Hemp house designed by Claude & Catherine Eichwald using lightweight frame system.

Pre-casting

Top; Hemp clay block
Above; Hempcrete panels cast with organic fibre rich hurds. Hemp/clay bricks used to repair cob structures.

Whenever the concept of making bricks from clay or mud was first tried, we cannot be sure, but structures such as the arches of ancient culverts made from this material have been found dating back 7,000 years or more.

For various reasons it was found to be more convenient to make bricks than to mould the wall in one piece. With a material such as hempcrete there will be situations where a hemp-lime block will be a more suitable way to build, than to cast an entire wall in one go.

The advantages of a hempcrete block are that the block can be many times the size of its clay or concrete equivalent. This of course speeds up considerably the time spent laying a wall and it removes the necessity of purchasing or manouvering shuttering, especially if it would entail working in situations where access was restricted.

When casting blocks from hempcrete it must be remembered that the material itself is much heavier when it is wet and fragile as it has yet to set. The hemp blocks are better left to dry before moving them.

Once it has lost around 70% of its moisture content hempcrete is quite suitable, so long as handling is kept to a minimum as it is too fragile to to be thrown about. The setting time is a factor that requires the blocks to be made at least a month or two in advance.

The manner in which the blocks are stacked during this drying time will need to ensure they do not adhere to each other. This is best acheived with a mesh layer between.

The design of the block itself will vary to some extent with the dimensions required and the conditions provided for the curing process. Some versions I have seen have been made with spaces or holes through the block in order to provide an aperture for pipes or conduits to pass through, however in my experience of building it would be very difficult at times to attempt to line up such holes to enable any such services to be fed through at a later stage. It is easier to drill holes where and when you need them. Holes in the blocks created when they are being cast do however, help the blocks to dry out more quickly.

Ceiling panels, which can be used between exposed rafters, are manufactured presently. The panels are made with hemp hurd processed in order to contain a percentage of fibre, this acts as a reinforcement in the matrix of the hempcrete, which enable the panels to be cast in thicknesses down to 50 mm.

My own experiments have shown that quite complex shapes can be formed from the hempcrete with the right type of mould, and this area has great scope for development, the fibres also help keep the cast shapes in one piece whilst they are still fresh.

This mix of fibres and hurds must be kept to a certain balance, as too much fibre in the mix tends to form in clumps and this causes a problem with moisture. The

Hemp blocks

lime and fibre mix in these clumps is too lime rich, due to the low surface area of the fibres. These fibres also act as a wick drawing the moisture into the hempcrete and keeping it wet for too long.

The ideal proportions of fibre/hurds for the reinforcing element to work correctly is between 5-15% fibre of the volume of hurd.

To use hemp blocks effectively a different kind of timber frame is utilised. To avoid having to position the frame actually in the hempcrete wall, an inner frame is erected to support the floors and roof. The hemp block walls are then built up around this frame with the doorways and window frames built into the walls as they rise.

To bond the blocks together a lime/sand mortar can be used and it is possible, with the right blocks and mixture of lime, to simply brush a jointing paste on to the blocks to produce a minimal joint with no thermal bridging.

Left; Hemp block cast with 50mm thick walls
Bottom right; House being built with hemp-lime blocks around independant frame.
Bottom left; Variety of hemp-lime blocks from trials by author.

Windows details

The creation of a window in the wall of a hemp building requires an effective method of shedding water at the lintel and at the sill. The style will of course depend on local climatic conditions; prevailing winds, rain, etc.

From a design point of view it is a good idea to think of the windows of a house as the 'eyes' and their dimensions and style will greatly affect not only the appearence of the building but the ambiance of the interior.

Here are illustrated two different systems, one suitable for an environment where the window is protected from the elements by mounting it in the inner frame of a thick cast hemp wall, the other a Tibetan example where a prefabricated window 'box' is set into the earth wall as it is built. Though not traditionally used with hemp, the concept could be adapted as the material could be cast around it in the same manner as earth would be. The Tibetan climate probably does not include high levels of rainfall, but any rain or snow is shed at the top by the canopy.

Below left; Wall built using double frame system with window set into internal section.
Below right; Tibetan farmhouse window built as framed unit set in cast earth wall.

Door details

The detailing of a window or doorway is a very important aspect of the design of a building. With the hemp system it will be important to finish the hempcrete at this position in the structure in a manner not only suitable to the climatic conditions, but also to those of the skills of the craftspeople working on the job.

The frame needs to be strong enough to take the constant opening and closing of the door, and so will need to be rigid enough for this, and the finishing of the reveals to the doorway need to be thought out to function correctly in order to shed rainwater or protect the corner of the hempcrete walls.

Several ways to do this can be employed with a number of different frame systems. The door may be set in an external frame so that the reveals are created by the timber, or the shuttering can be made to produce a hempcrete reveal which can be rendered in the normal manner.

Above; Doorway with timber board as reveal on unrendered hemp wall.
Left; Doorway with rendered reveals

Lighting

To bring out the unique visual attributes of a room created using hempcrete, it is best to position the light sources so that they shine across the surface of the material to highlight literally the texture of the material.

This can be best achieved by either building a light into the surface of the hemp or by placing a free standing lamp in a suitable position.

Left; Table lamp hightlighting texture of the hemp plastered wall.
Above; Light fixture built into the surface of the wall like a swallows nest.

Kitchens & Bathrooms

Left, Cosy bathroom plastered with hemp.
Below; Decorative tile mosaic behind a kitchen sink.

In situations such as bathrooms and kitchens where there is likely to be a higher level of humidity than other rooms in the house, hempcrete works very well with the distribution of moisture and prevention of condensation.

On areas such as the splashbacks of sinks or baths, tiles can be used to give an easily cleaned surface, provided a breathing lime mortar is used as adhesive and grout.

RENOVATION

Cottages

The properties of hempcrete previously described, can equally be applied to the renovation of existing buildings, as it can to new structures.

The differing types of hempcrete can be utilised in the same situations, the only difference being the installation methods or techniques. These will vary due to the fact that the material is being applied to an existing building and could entail climbing into cramped spaces or protecting furniture or decorations whilst working in an inhabited room.

Cottages come in all shapes and sizes of course and can be classified as cabin or farmhouse. They might be situated on exposed coastlines or in sheltered forests and all these factors will have to be considered when approaching such a project.

The easiest way to categorise differing types of cottage is by the materials they were originally constructed with. In this way we can specify how a particular type of roof or wall etc, might be approached.

Many old buildings are in need of an upgrade when it comes to the comfort we now expect from our homes and it is in this area that hemp really comes into its own. Most old houses or cottages are built using local or traditional building methods and have structures that breathe, so it is essential that they are repaired in a fashion that works in the same way.

Opposite; Hemp plaster on stone walls and partition.
Right; Lime plaster on sprayed hempcrete insulation.

Thatch Roofs

Top; New straw thatch.
Bottom; Peat layer laid on roof structure beneath thatch.

Whenever we see a thatched cottage, we think how warm it must be tucked up under that cozy lid. However, there are some drawbacks, the first one most people think of is of course fire.

In reality, on a day to day basis, it is more often that they can be quite draughty, even with a straw thatch which allows less air to penetrate being more tightly packed together than reed. This draft may be sealed off behind a surface of boards beneath the rafters, but the space between will be full of moving air on a windy day which reduces any insulation properties.

The modern solution to the problem of fire is to use a foil membrane beneath the thatch, unsupported or on a layer of plywood or O.S.B. sheeting. Unfortunately, this is seldom satisfactory as in the process of their work the thatchers puncture the foil (which is only marginally thicker than that used for baking) as they probe the thatch with the crook feeling for the rafter to fix into. Moreover the foil prevents the movement of moisture causing more problems with the loss of breathability.

Fire is mostly seen as the greatest risk for a thatch roof, even though the statistics don't back this up. The bad reputation was established with fires, such as the Great Fire of London in 1666, and reinforced with the coming of the spark spewing steam powered traction engines amongst the thatch farmsteads of the 19th century. Early attempts to create a fireproof solution were rediscovered whilst researching archives for the rebuilding of the Globe theatre in London. It was found that Londoner's used to paint their thatched roofs with limewash, which must of created a vista of the city under a permanent layer of dirty snow.

Methods of thatching vary as you move westwards across Europe. In more eastern areas systems were developed where the straw or reed was tied with twine or pinned with iron crooks to the rafters. More recently, thatchers might secure the sways (the lateral rods which the thatching material is held down with) using wire fittings screwed to the rafters.

The oldest method still used in Ireland and Scotland up until 60 years ago was that of fixing the thatch with spars driven into a layer of bog turf. This material called the 'scraw' in Ireland consists of the top layer of the bog, the grasses, heathers and roots, stripped off prior to peat cutting for fuel. This was cut into large turves and placed on the roof structure and used to peg the thatch into. It provided not only a means of securing the thatch, but also a draught excluder with added insulation. This can be reproduced with a fireproof, breathing and insulating layer of hempcrete.

Extensions

Extensions are included here, as very often when renovating an existing building they are built as an addition to provide extra room.

This is especially true where more light is needed in rooms such as kitchens etc, by using a timberframe it is easy to install larger windows than might have been possible in existing walls.

Once again the hemp building system lends itself to this type of project very well being extremely useful when it comes to 'blending in', also the textures obtainable with hemp compliment those of a stone wall or old beam very well.

The material is ideally suited to marrying two parts of a building together being flexible in its structure and the variety of finishes that can be applied.

Below left; Terracotta block extension, with hemp plaster, onto stone house.
Below; Hemp extension to old cottage by Ralph Carpenter,architect, England.

Castles

When approaching the restoration of a building such as a castle there are obvious differences of scale and size of the components of the structure, which in turn affect the way those materials work. Other differences to the job of repair will be that historical aspects of the appearence and decoration of the castle will have to be maintained, whilst at the same time damage rectified and the whole made habitable in the modern sense of the word.

The main difference between a castle and other habitations is the thickness of the walls. Due to to their height they have to be wide enough to be stable, nevermind the defensive considerations. This mass of stone has great potential for heat storage, and in the days in which these edifices were built they were surrounded by great forests which supplied the fuel for the many fireplaces found within. To be able to utilise the walls in this way today would entail insulating the exterior of the building, which would seriously affect the appearence and would go against the primary aims of restoring a castle.

What can be done is to install a breathing insulation to the interior walls, in order to shorten the heat up time of often large internal spaces, and to find some way of unobtrusively bringing heat into the building. To do this in a way that works with the fabric of the building and the often priceless antiqities kept within, the system of running what would normally be underfloor heating pipes on a hemp insulated wall beneath a traditional lime plaster, creates an environment that has an authentic appearence with contemporary comfort.

Opposite; Chateau stairway with heating pipes fixed on hempcrete insulation, covered in lime plaster.

Top; Towerhouse with lime render.
Middle; Close up of towerhouse wall
Below; Chateau room with hempcrete insulation behind heating pipes finished with lime plaster.

HEMP
PLASTER

Origins

To use the term "origins" might be claiming a bit too much, in that the actual use of hemp plaster wasn't made until hempcrete was developed and its properties discovered.

We could say that a precedence was established in vernacular building much earlier, in the use of daubs applied to wattle structures or the use of hay mixed in a clay plaster sometimes used to help insulate old farm dwellings for instance.

Many proportions of the constituents of the mix have been tried to suit various application techniques. From those applied by hand, to those sprayed with compressed air guns. Some methods suit only hemp & lime, others need the addition of ground limestone and sand to help retain moisture during the setting period, and to give the material enough weight to work effectively in a spray gun so that they can be applied at greater speeds.

These variations affect the appearence and performance of the plaster, but they all provide a great improvement to the comfort of the living spaces they are applied to.

It is this utilisation of hemp and lime as a plaster that has been the simplest way for self-builders to include it as an experiment in their own projects, and in Ireland alone there are over 100 such examples.

Below; Hay/clay plaster.

Opposite; A combination of textures attainable using hemp plaster.

Old Buildings

The application of conventional plaster in the renovation of existing walls can be a tricky business, due either to the necessity of appyling many layers to achieve the required thickness, or because there are such a variety of materials in the background to be plastered.

In these kind of situations, hemp plaster is ideal as it can be applied at a thickness sufficient to level out any unwanted lumps on a surface and will adhere to many types of background.

Some hemp producers are separating the finer particles of hurd to provide a more suitable material for plastering. This produces a very smooth surface especially suited to application by trowel.

I have found that many people attempting their own building projects especially enjoy mastering the technique of working with this tactile material. Moreover, the effect of a quicker response to a heating source that the plaster gives to the occupants of an old building, compared to what they remember of the old cold days, makes it even more rewarding.

Opposite; Cottage plastered with fine plaster grade hemp.
Below left; Close up of window detail.
Right; La Maison Diocésaine, Châlons-en-Champagne, France, modernised as office space with hemp plaster sprayed on interior of external walls.

New Buildings

Although there are many benefits to living in a modern home, some people are dissatisfied with the box like feeling to the rooms in such a house. This is a situation in which hemp plaster can be used to great effect, to soften the lines and give texture to what would normally be a flat surface.

With this application the insulating properties of hemp plaster are not the main consideration. It is primarily the visual effect of the material in changing the way light reflects from a wall or the way edges might be softened.

It might also be utilised to give the feeling of an older building to a new one, or to blend older sections into new additions.

Opposite; Newly built house with hemp plaster to create 'old' feel.
Right; Modern house of terracotta blocks using hemp plaster to soften edges.
Below; Gallery with hemp plaster to compliment ceramic art.

Textures

One of the most utilised aspects of hemp plaster is its texture.

The manner in which light is reflected off the surface of a wall will vary tremendously if the wall is plastered with hemp, whether it is bright sunlight or candle power, it will certainly provide character and add a uniqueness to the space.

Applied to a rough stone wall it will soften all the rough shapes and used on a flat wall can be used to add interest.

1; Rough hemp plaster painted with thick limewash.
2; Rubbed finish.
3; Rough unfinished.
4; Sprayed hemp plaster finished with wood float.
5; Smooth hemp plaster limewashed.
Below; Textured hemp plaster on stairway wall.

Opposite; Painted textured hemp plaster

Getting Artistic

As hemp plaster is such a malleable material it can be used to create quite artistic effects.

If the mixture is made into a stiff creamy consistency it can be moulded and shaped into relief decorations, either as you apply the plaster or later as it is drying out.

The surface of a hemp plaster if used with a minimum of lime is also a suitable background to apply coloured lime plaster designs, providing plenty of texture to key into.

Remember that if you create shapes in relief on the surface of a wall that they will be likely to collect dust if you don't shape the upper edges to prevent this.

Above; Hemp plaster moulded to form a tree around a window.
Right; The rough plaster finish is ideal to apply raised plaster motifs.

External Hemp Render

Hemp plaster is not generally used externally, but occaisionally there are situations where its qualities are ideal. This is especially the case with strawbale houses.

Normally these type of buildings would be covered externally with expanded steel mesh to give a lime render purchase on the weak background of the bales.

The use of hemp plaster removes the need for this, as it is a much lighter material and can bond with the straw surface without this support.

If this is then painted with several coats of limewash it gives a practical and more appropriate finish to the exterior of such a building.

If the insulation qualities of hempcrete are needed on the exterior of a masonary building, then it is better to use the wall mix cast, or sprayed, against the wall and finished with a lime render. This will insulate the wall allowing the stone wall within to act as a thermal mass.

Right; Close up of external hemp plaster.
Below; Strawbale house with external hemp plaster.

FINISHES

Cast walls

The surface of a cast hempcrete wall can look very attractive, as the colour of the tannin in the hemp comes through the lime to a greater extent than it does with hemp plaster.

The practicality of this will vary depending on aspects of how it was actually cast. The manner in which the hempcrete material is compressed when packing it down in between the shuttering or the surface of the shuttering itself, will affect how well the surface will stand up to the wear and tear of everyday living.

Alternatively, the wall can be treated with a lime paste of the same proportions as the lime mix in the hempcrete. This can be painted on as a thick limewash to fill the hollows and bind the surface particles together.

Opposite; Cast hemp wall finished with thick limewash.
Left; Bottle set into hemp wall finished with limewash.
Below left; Untreated cast hempcrete wall.
Below; Cast wall highlighting strata of hempcrete.

Moulded features

By first fixing a shelf or feature made from timber or other material to the wall, it can then be covered over with the hemp plaster to become a useful moulded feature in the room.

Right & below; Moulded shelves.

Exposing

In some situations it might be valid, artistically, to expose the underlying composition of the wall, be it a hemp wall or the stone wall beneath. This can be both visually attractive and practical as in the exposing of stone at the base of a doorway to better take the knocks of everyday use

In another use illustrated here, the original plaster was left on the wall because if was too difficult to remove and so the shape was defined by the hemp plaster.

Left top; Exposed hempcrete leaf.
Left bottom; Exposed stone.
Below; Hemp plaster framing original plaster.

Rubbed

An attractive and practical way of finishing the surface of a hemp plaster is to rub the surface with glove clad fingers to embed the hurd fibres into the lime creating a smooth appearence to the surface.

If you are plastering a wall that is not flat or square with sharp corners, such as a cob wall moulded by hand, or a masonry wall that has been built to have those type of characteristics of curved lines, then this type of application can be used to accentuate that.

This produces a surface that is suited to an unpainted finish or will mean that you have less surface area if you do decide to paint it.

Left; Rubbed hemp plaster on alcove

Trowelled

As with any plaster type material the way in which it is applied will affect the finished appearence.

Finished in this way hemp plaster applied with a trowel and only partially smoothed over will provide yet another type of surface. The variety of textures possible is very wide, but if a trowel is employed it is better to finish it with a lime paste to smooth over any impractical edges.

Right; Hemp plaster applied with a steel trowel and smoothed flat later.

Painting

Produced now in a range of colours not available in the past, lime paints are now coming back into vogue. They are of course, being made with lime, perfectly compatible with hempcrete surfaces, meshing with it and behaving in the same way as regards humidity and moisture.

The difference between old types of paints, whether lime or oil-based and the modern plastic-based products available is best put in this way; petrochemical paints, once they are applied, continue to dry by evaporation of their spirit solvent until they are so dry they crack off. While the older types of paint never really dry out, especially lime-based paints which always hold a certain amount of moisture thus keeping them flexible and adhered to the surface to which they were applied.

Lime paints reflect light in a unique way due to the shape of the minute particles they contain and work very well with the often textured surface of a hemp plaster.

Other mixtures of lime and casein or quark were developed in the past and some of them will produce interesting textures when applied to a hemp plastered wall or the flat lime rendered surface in a hemp building.

Opposite; Lime paint on rough plaster
Above; Selection of lime paints
Below; Watered down lime paint on hemp plaster.

Internal Board

The use of timber sheeting as a surface to the interior walls of a house built with hempcrete is a simple way of finishing this type of structure. In place of having to move shuttering as you build, boards such as chipboards or O.S.B can be fixed in place to form the finished surface and painted at a later date.

They provide a cheap practical solution to this facet of design with hemp building systems, in that it involves less labour in erecting and removing shuttering and adds a certain element of added strength to the timberframe. Of course if the cost of timber used to make this kind of board increases to a point where it could be economical to be replaced with hemp, it would provide another role for the plant in building.

The components of hurd and fibre which could be used to make such a board have been shown in tests to adhere to each other well, without the addition of glues, when used fresh and green. Whether this could be manufactured in a form to satisfy durabiliy requirements remains to be seen.

Left; Orientated strand board used as internal surface with painted finish.

Cloth & Paper

To cover a raw hempcrete wall without using plaster it is possible to utilise cloth or paper.

If ironed and then stretched across areas of hemp wall, cloth can be painted with breathing paints in the same way as an artists canvas. Anything from a rough canvas to fine linen can be used, depending on the surface texture you are looking for, it is then only a matter of deciding on the way you will manage to join the sections of cloth. This could be by neatly fixing with pins, which are left exposed or hidden under a moulded beading.

The use of cloth in this way reduces the amount of water used in the construction process, this has many advantages especially with drying out times in certain climates.

Alternately, if the surface of the hempcrete is perfectly flat, wallpapers could be employed. It is probably preferable if a layer of lining paper is applied first to provide a flatter surface to apply the final paper layer. If the rough texture of the wall is required it can still be covered with a layer of tissue paper, which is available in many colours, and this can be pasted to the surface in whole sheets or as a patchwork of torn pieces of different colours in the manner of papier-mâché, as in the picture on the right.

Above; Cast hemp walls with lime render

Lime Renders

The traditional weatherproof covering to the exterior walls of a building can be either timber or a lime render. With the trend towards more organic looking buildings promoted by books such as the one you're reading now, the use of lime render is becoming popular again.

Of course the appearence of a plaster or render is due entirely to the skill of the craftsperson/artisan who is doing the work. As most plasterers have been trained to be able to execute a perfectly flat finish to their work, they may look at you as if you have two heads when you ask them to make it not so flat. After all there is only one version of flat, but endless variety when it comes to 'rough'.

You will probably need to show them what you mean by rough or find somone who has done that kind of work before.

Right; Smooth lime render finish to house with hempcrete insulation to internal walls.

Below; Sprayed lime render finish to stone building with hemp plaster internally.

Boards

The use of wooden boards to clad the exterior of a timber frame building is a natural choice when the primary skills used in the building are those of the carpenter. None of the wet trades such as plasterers need be involved.

Normally softwoods are employed, fixed to the frame horizontally and overlapping to shed the rain, hence the term lap-boards (in some areas they are termed clapboard).

These boards were then traditionally painted with a limewash in some cases, or bitumen or tar in others, to preserve them from the elements.

Recently it has become the fashion to use timber boards vertically one layer overlapping the first to seal the joints. These can then be painted with a protective coat of oil, which also brings out the colour of the wood.

Below left; Old timber clad house showing details of the porch and windows.
Below right; New house with vertical larch boarding painted with oil.

Boards

It is possible nowadays to purchase pre-painted boards manufactured with profiles to be used as exterior cladding.

These together with prefabricated timber frameworks can provide a quick way of erecting the shell of a house, to then have hempcrete applied to the interior.

Provided of course the boards are painted with a breathable paint, the hemp can be sprayed or cast against the exterior wall once the pumbing and electrical first fix has been installed.

Below; Timber frame house with sprayed hempcrete internal wall finished with lime plaster with external lap boarding, France.

Tile & Slate Hanging

Below; Glass and steel structures can be made more comfortable with the use of hempcrete in floors,roofs and wall areas.
Below right; Slate hung wall facing.

Tiles and or slates have Long been used as a way of cladding timber framed buildings and exposed stone walls, they provide the structure with a mineral covering as an alternative to boards or plaster.

Fixed to battens on the exterior of the wall with nails or screws they were also traditionally given the additional support of adhereing to a lime plaster applied between the battens on a stone wall. This could be replaced with hempcrete in order to provide some element of insulation on the exterior of a stone building

They are suitable as a covering for hempcrete walls as they will allow the surface of the hempcrete to breathe and prevent any rain from entering the building fabric surface.

Shingles

Choosing to clad the exterior of your building with shingles will largely be influenced by where you live and the traditional use or not, of such materials.

Being made of wood, shingles are a more breathable material than slate but of course more likely to decompose if not treated or protected from the elements.

Usually made from cedar or chestnut they can be made from any timber that splits easily to form these tile-like pieces. In Switzerland small waste logs are utilised in this manner and the resulting little tiles are called "dragons teeth".

Above; Use of slate hanging with other materials provides a attractive finish that would be suitable for the hemp building system.
Right; Creative use of shingles.

Cordwood

The term cordwood refers to the method of measuring logs for fuel using a cord, a cord being 128 sq feet of fire wood. However in this case we are talking about the technique of using short sawn logs stacked to form a wall. The sawn ends forming the inner and outer surfaces.

Formally these logs were bound together with a mortar made from lime and sawdust, but the sawing of sufficient logs for the building does not produce enough material for this and a hempcrete mortar works better than one using sawdust in any case.

The walls of such a structure can be built quite quickly, though to achieve a flat surface to the walls the logs must be cut to precise lengths.

There are many opinions as to which type of timber is most suited to be used in cordwood buildings, the main aspect seems to be that the timber is well seasoned so as not to shrink away from the mortar.

The surface of the cordwood wall will need to be pointed between the logs and the whole thing can then be painted using breathable paints or left as they are to show the natural colour of the timber.

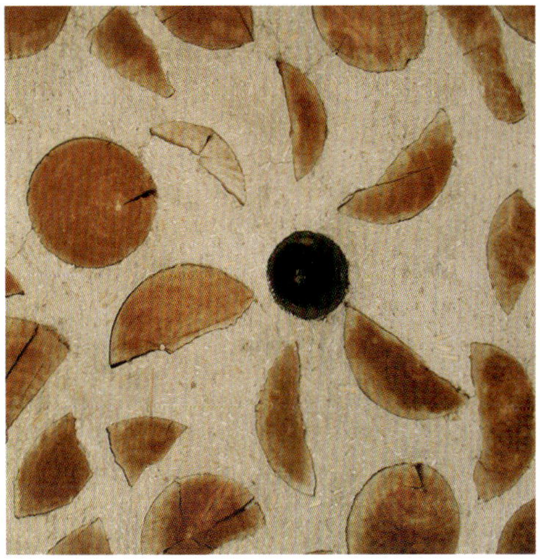

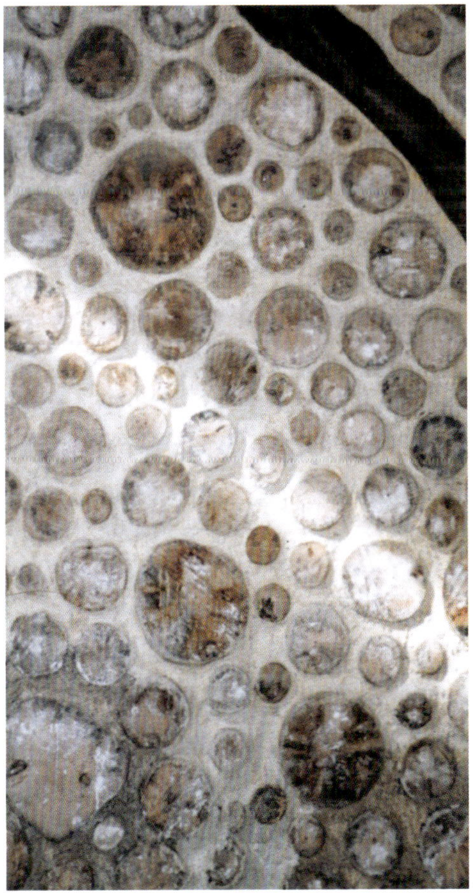

Opposite; Cordwood cabins, Provence, France. Inset; interior of cabins.
Left; Cordwood wall in process of being pointed with lime.
Above right; Detail of cordwood wall with bottle added.
Right; Painted cordwood wall.

Hemp fibres &
Other materials

Hemp fibre & Other materials

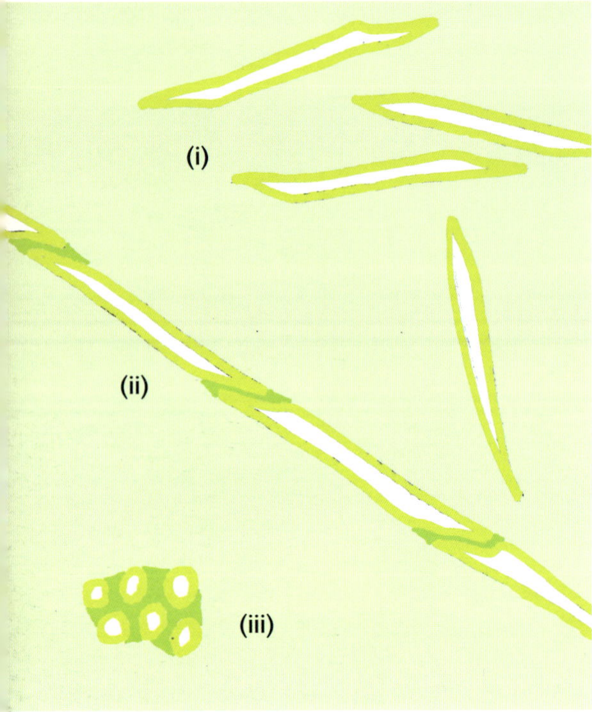

The fibres contained in the skin or bast of the hemp plants stem are considered to be the longest and strongest of all the plant fibres.

Whilst the plant is growing they surround the less flexible wood core and reinforce the structure of the stem so that it is able to withstand the wind.

They consist of almost pure cellulose and are in reality short fibres of 20-30 mm long, joined together with hemicellulose and pectins to form long fibres some over a metre in length. These fibres are formed in bundles within the bast and are themselves surrounded by pectins when growing, this then turns to lignin as the plant matures and starts to produce seed.

The presence of these fibres around the stem cause most of the problems associated with the processing of hemp, but were also regarded as more valuable than the hurds were in the past.

However with the switch to short fibre spinning and use in modern textiles, hemp fibres are now primarily used as reinforcement and matting in the automotive trade in Europe and for speciality papers, such as for banknotes and cigarette papers.

In China the hemp textile industry is still operating with a high output for both their domestic market and for export to the West. In the area of hemp construction the fibres are used primarily to manufacture insulation. The hollow fibres provide an extra element of trapped air to give hemp fibres an advantage over solid cellulose fibres. As hemp fibres are cellulose and not protein, as is wool, moths are not interested in consuming them for food and so they need no chemical or other treatment to prevent this.

Above; Diagram of hemp fibres showing (i) individual fibre segments (ii) segments joined with pectins to form long fibre (iii) cross section of fibre bundles.
Right; Loose hemp fibres used for insulation

Insulation Panels

It is possible to use hemp fibre treated with a fire-retarder and inhibitors just as it is, stuffing it into roof or loft spaces. However to achieve a regulated and quantifiable level of insulation it is incorporated with other materials into a more acceptable form of building product.

The fibres are coated with usually borax and boric acid as fire preventors and mixed with polypropylene fibres on a moving bed. This is then moved through an oven to slightly melt the plastic fibres so that they stick to the hemp fibres and each other, and bind the materials into a matted structure.

The different manufacturers produce their products by various slightly different methods and the layers of fibres are fed into the machinery so that they form either flat stratas of fibres or they are pleated which gives them slightly more resistance to compression over time from the forces of gravity.

The mats are then cut into flat panels or batts or packed as rolls to be layed out and cut into strips to suit the particular job.

To cut the insulation to size on site, a serrated knife similar to that used to cut bread is employed.

Top Right; Organic hemp fibre insulation.
Middle Right; German hemp insulation.
Bottom Right; Pleated hemp fibre insulation.
Below; Hemp insulation batts on site

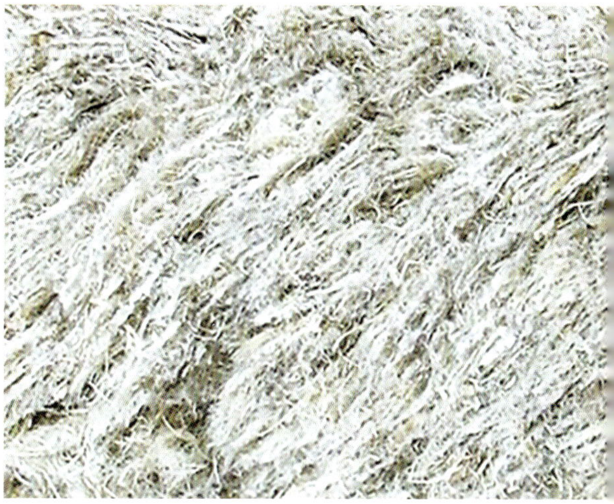

Carpet & Underlays

Carpet sales have fallen recently due to the increase in people that suffer from allergies.

The dust mite seems to be largely responsilble for these allergic reactions and carpets provide a place for these creatures to dwell in larger numbers than in the past. This is probably because we use carpeting in different ways than we did previously.

In the past carpets were not fixed down as fitted carpets are nowadays and before the invention of the vacuum cleaner, were rolled up and taken outside to be hung on the clothes line and beaten with a special beater until the dust was removed.

Despite this many people still like to have carpeting beneath their feet and hemp can be used to make hardwearing carpets or rugs.

Other fibres such as jute are still cheaper than hemp, so large scale use of hemp in carpet manufacture has not been taken on as yet, however small scale production in India has been started and large firms have done research with the material.

Carpet underlay made from hemp fibres is produced in France and this can also be employed as a sound proofing layer beneath floating timberfloors.

Left & Below; Indian hemp rugs
Left corner; Hemp fibre carpet underlay.

Canvas & Rope

Twisted hemp fibre is an early example of a hemp product. The rope and twines produced by primative peoples out of which they fashioned their clothes and other items have evolved to the highly technical industry of modern textiles.

Today it is not possible to find real canvas (cloth of cannabis) as was made for the sailing ships of yore, but softer versions used in the present ecological fashion market are available. These can be used to not only line a wall, but as upholstery cloth or other furnishing fabric such as curtains.

On small building projects the builders may still use a bucket and rope on occasion and if this is the case it is still possible in certain regions to buy a length of hemp rope for this purpose. As any builder will tell you, such a rope is far more easier on the hands from day to day.

Hemp rope could also be used to substitute steel cables in a structural system that relies on an element of tensioning for its stability.

Twines made from hemp are also still produced and these have many uses around the home and garden and could be utilised in the area of construction if lashing was to be employed as a way of securing components to one another as was done in the ecologically correct distant past.

Top; Canvas.
Middle; Hemp rope.
Bottom; Hemp twine.

Terracotta

Left; Mixing shades of colour.
Above; Selection of terracotta tiles.
Below; Chateau floor with newly laid terracotta tiles.

Terracotta

There are two main reasons to use terracotta products in a hemp building.

The first would be, in the case of floor tiles, the need to use a breathing material especially in this position on the floor where there is no plastic waterproof membrane. The second is the energy consumed in the manufacture of such products.

The industrial production of tiles is of course carried out in very large factory installations these days and because of this the fuel consumption is carefully calculated to be as economically viable as possible. However, the difference between the temperatures required to fire glazed waterproof tiles at 1200-1300 degrees centigrade, compared to 1000-1080 degrees centigrade to fire terracotta, are considerable. It is even possible to obtain terracotta tiles that are baked by the sun alone which would be a much lower energy consumption. These tiles are not as hard wearing as those fired at higher temperatures but are ideal for areas of low traffic.

Drainage and sewerage pipes are also made from clay to meet the needs of those builders trying to avoid the use of petrochemical products. These are quite fragile and need to be handled and installed with care as stones falling into trenches with uncovered pipes of this sort can cause cracks and breakages all too easily. They can be joined together with plastic sleeves as these allow a greater amount of flexibitilty in the line taken by the pipes in a trench than would be possible with a male/female joint of terracotta.

Top; Terracotta pipes joined with plastic sleeve.
Bottom; Selection of terracotta pipes and joints.

Hempcrete Additives

In addition to purely hemp, limes and cements, other materials can be added to hempcrete to make it more suitable for particular applications or to change its' appearance.

To give hemp plaster colour throughout the material, pigments may be added. These can be natural mineral colours or man-made pigments such as acrylic ones used to colour printing inks. Mineral pigments tend to come as powders and should be mixed with water first and left to stand overnight in order to mix with the hempcrete without forming spots of pure undiluted pigment. Acrylic colours usually come as a liquid and should be diluted with water before being used.

To give hempcrete a harder surface or more thermal mass (but less insulation) sand or gravel can be added to the mix. If the hempcrete is to be exposed without a covering of plaster, remember that the colour of the sand will affect the appearence of the hempcrete.

Some situations in a building will not be suitable for the use of hempcrete, such as under the floor of a "wetroom" type of shower. In these situations it is better to use a material such as expanded clay which is lightweight but still breathes and can be used with a limecrete screed on the top of it.

Other materials added to the ingredients of hempcrete can be used to replace or supplement the basic ones, such as lime or cement. Pozzolans will speed up the setting time of the limes (they are available as ground volcanic ash from Pozzuoli in Italy) or crushed brick dust. They work by drawing carbon dioxide into the mix quicker than it would be normally.

Materials such as calcined clay, made from burnt china clay, will give the binder element of the hempcrete higher density as the particles are smaller than some hydraulic limes.

Above; Expanded clay pellets.
Right; Mineral pigments.

Roofing products

The problem of disposing of used vehicle tyres will be with us for some time yet, but it is always good to hear of a way of recycling them as a useful product.

The tyre rubber is removed from the steel and plastic reinforcing meshes and formed into granules then melted together with plastic and chemical fire-retardents to be moulded as roofing tiles.

They are injection moulded to the shape and profile of traditional slates with rough edges and surface and offer the builder several advantages. As they are tough and flexible they can be dropped or stood on when on the roof without breaking and are cut with a standard retractable blade knife. Holes can be drilled in many tiles at one time without cracking any of them and they can be fixed using a nail gun.

These factors all add up to less time spent on labour to roof the building.

Top; European recycled rubber slates
Bottom; Canadian recycled rubber slates.

HOW TO

Site Organisation

A building site can be a dangerous place no matter what the materials used to build it are made of. The safety of all who work on a building site must be taken into account when planning any project.

All equipment on site from electricity generators to mixers must be installed and operated by qualified personnel. Any scaffolding used for particular jobs must be erected and maintained also by those trained to do so.

Factors to be considered when building with hemp-lime are those such as storage of materials on site and their position with regards to lifting or transporting them to be used. The position of the stacks of hemp bales and pallets of lime sacks should keep movement to a minimum both to prevent injury and for reasons of efficency.

Water supply to building operations should be kept seperate from drinking supplies and electricty.

Timber should be kept dry and remember that it can also leave large or small but equally painful splinters in your hands so, when handling sections of frames, joists or beams, wear gloves!

To lift large sections of timber framework it might be necessary to operate a crane or provide access to a mobile telescopic lift. Ensure that all points around the structure can be reached safely from where they are to be positioned.

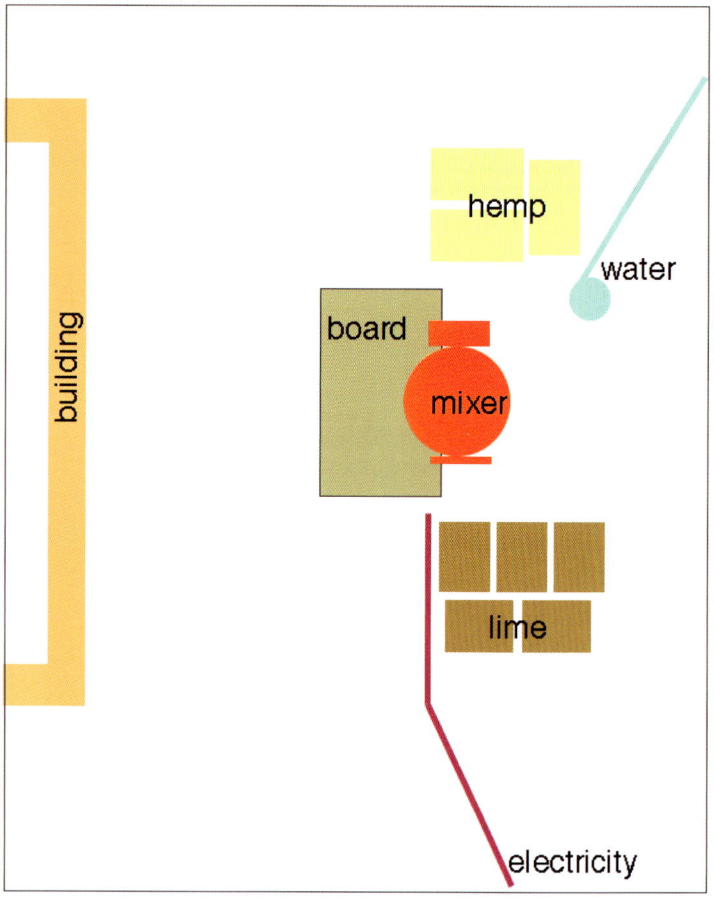

Diagram of site layout.

Site Organisation

As the majority of materials used in the system of hemp building are biodegradable, they in themselves do not leave behind any toxic or damaging residue.The hemp hurds provide a quickly composted mulch if they are spilled anywhere and if mixed with lime, will only raise the 'ph' of the ground they are disposed of upon.

The ground around the place where the mixing was carried out will have lime plaster residue as well as hemp and lime and this can either be utilised as a base for the drive or path to the house, or a part of the garden for lime loving plants to be grown in.

If the site is exposed to strong winds it will be advisable to shield the mixing area to prevent too much dust flying about during the mixing of hempcrete. Or if this is not possible, then maybe wait until the wind has died down.

Top; Mixing hempcrete can be a dusty business.
Bottom; Waste hempcrete causes no problem for the surrounding environment.

Plant & Equipment

Building with hemp in most instances does not require any unusual equipment, but of the array of standard equipment used on a construction site there are some that are more suitable than others.

Most building projects involve the use of some form of scaffolding in order to access parts of the structure. With a hemp building, jobs such as erecting the framework might need this approach especially if the frame is to be built on site. However, a permanent scaffold surrounding the building can obstruct access at times and so often it is better to use a lightweight movable scaffolding tower system that can be erected where and when it is needed.

An alternative to this is to use a crane to lift heavy objects into place or to suspend a platform from which to work off. There are now small cranes available that are self-erecting and can be positioned on a site so that they can reach all parts of the structure during the building period, without the need to have a clear space on all sides of the structure. They are ideal for lifting sections of pre-fabricated frame into position, and if used with a hanging platform can be used to lift a worker into postion in order to spray hempcrete into areas that otherwise would need walkways and ladders erected, such as on the roof.

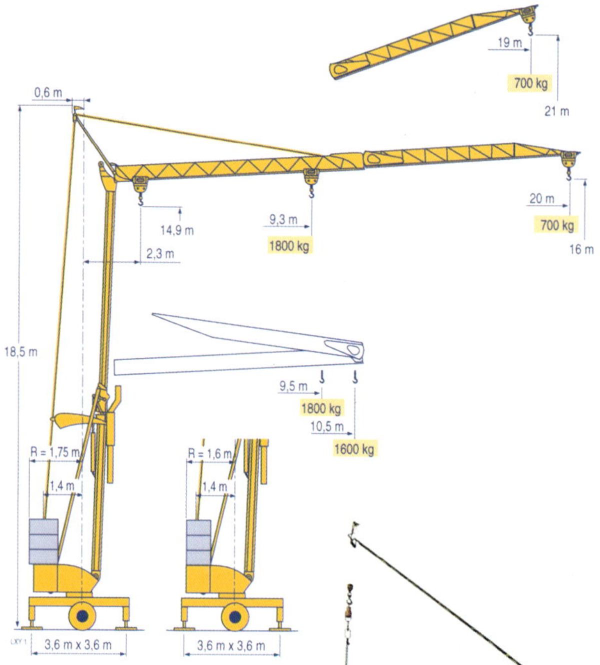

Self-erecting cranes.

Plant & Equipment

The alternative to using a crane is to use a piece of equipment that has become very popular on building sites these days, and that is the telescopic lift. These machines provide not only a method of lifting materials into position but also a more stable temporary platform to work off, if that is what is needed. However, it is also necessary to be able to access the building from all sides and so the ground would need to be fairly level and firm around the structure.

When it comes to using mixers for the hempcrete, it is better for the drum capacity to be as large as possible as the volume of hempcrete needed to build with is quite large and can be used up quickly. An electrically powered machine is preferable as it is more convenient to be able to stop the drum between loads and this is awkward with a diesel motor unless it has an electric starter motor.

When spraying hempcrete plaster you will need a diesel oil-powered air compressor, as most electrically powered units do not have sufficient capacity.

Together with the compressor required for the concrete pump, a small water tank and pump will be needed to supply water to the hose inlet. To power the mixer and concrete pump a 3 phase electricity generator will have to be provided.

Top; Mobile telescopic lift and large drum mixer.

Casting Tools

The tools needed for casting a hemp wall are split into two categories; those to fix and move the shuttering, and those to mix and then cast the hempcrete in place.

To secure the shuttering boards you will need a cordless drill to make holes in the boards and a hammer and spanner. A ratchet type socket spanner is the most efficient.

To fill the cavity with hempcrete a large lightweight shovel is best, such as those made from aluminium used for shovelling grain or snow. The hempcrete is light but bulky and this shovel can also be used to scrape the shuttering boards clean between lifts.

To compact the hempcrete between the shutters it is best to make your own tamping tool to suit the type of framework you are working with. It is better to keep the size of the head of such a tool small enough, so as to be able to exert sufficient force at each action and not dissipate it over too large an area. I have also found that it is a good idea to have a pointed end to get into awkward spots. To make the one seen in the illustration on the right you will need a length of broomstick or tool handle (long enough to reach inside to the base of the shuttering) and a piece of 50x 50 mm timber. Then simply drill a hole in the timber to take the handle and fix with some glue and a wedge from beneath.

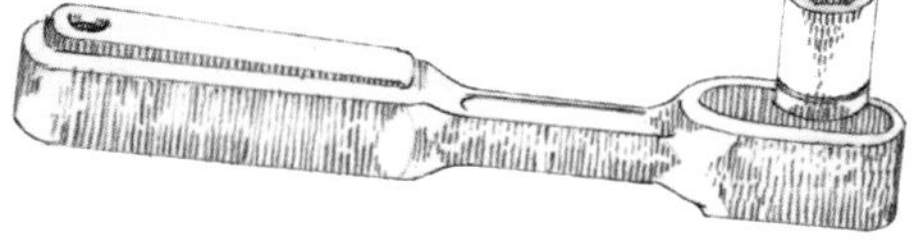

Socket spanner

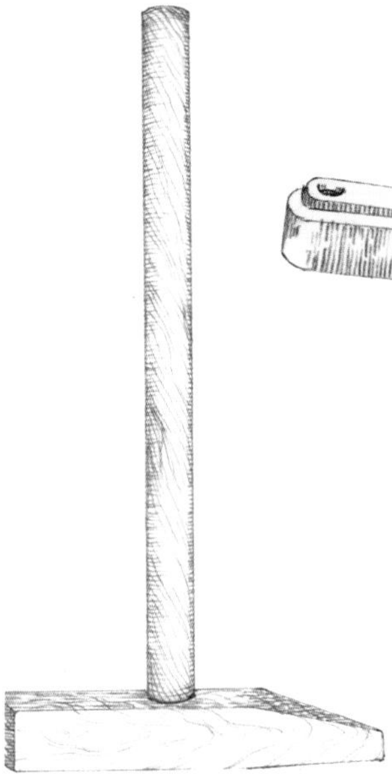

Tamper

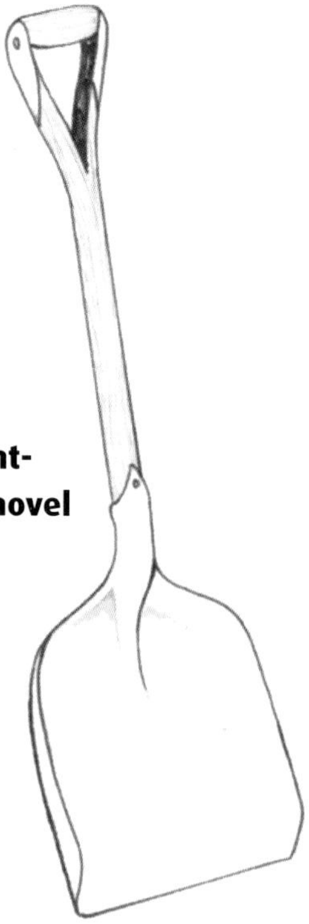

Large light-weight shovel

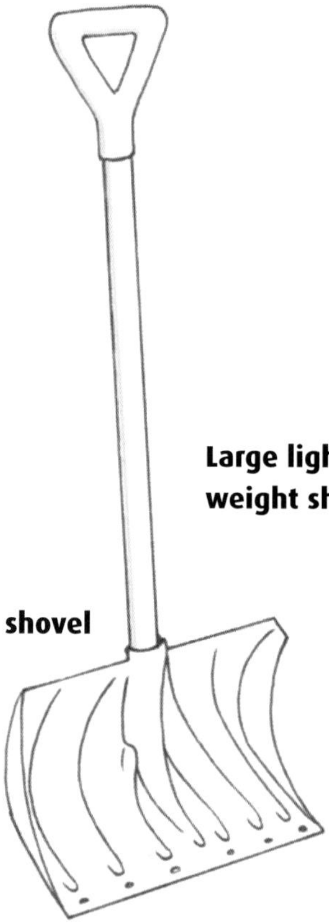

Snow shovel

Tools for Hemp plastering

Most of the tools used in hemp plastering are the same as those used with lime. Spot boards are set up near to the work at the right height so that the worker can easily take what they need without delays. The plasterer will take each trowelful off a hand held mortar board or hawk and apply it with a trowel or daubing by hand.
When applying with a trowel it is initialy best to use a steel float or trowel for the material to slide off the tool surface, but to work it smoother and flatter you will need a pointed wood or plastic float. The rectangular type are unsuitable as the corners catch the hemp particles and can cause the plaster to fall off.

Sponges and scrapers come in handy when finishing and cleaning up around a job.

If the hemp plastered surface is to be coated with lime plaster you will need lime plastering tools such as scratchers and specialised trowels for corners and hard to get to bits.

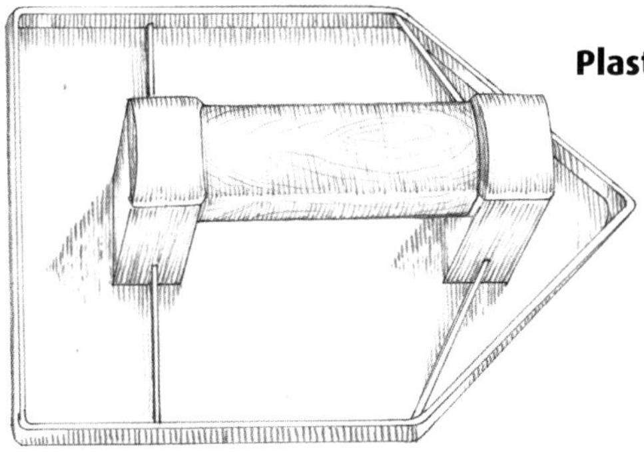

Plastic float

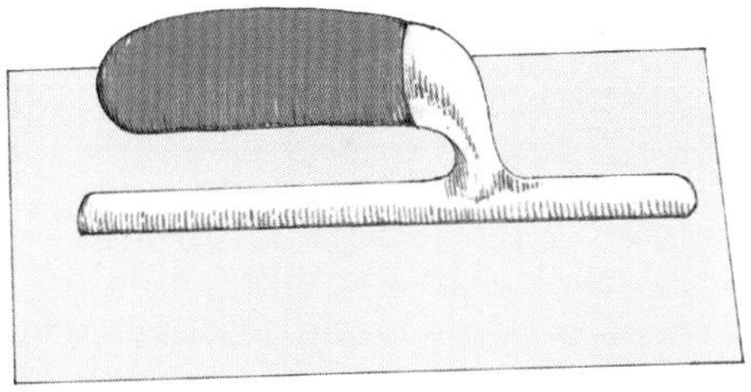

Steel float

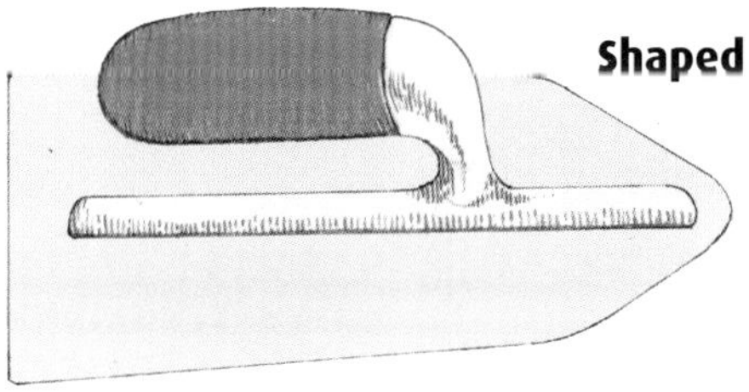

Shaped steel float

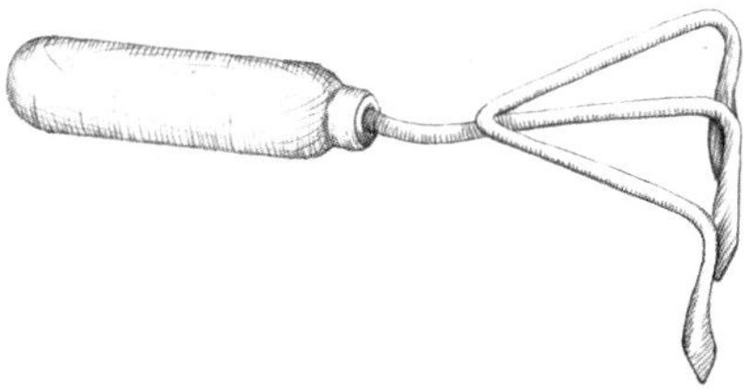

Scratcher

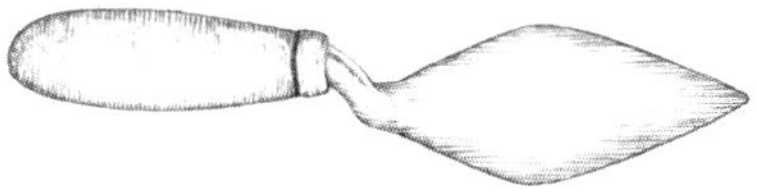

Small trowel

Protection [Casting]

The application of the hempcrete mixes used in floors, roofs or walls is a less messy business than hemp plastering, however it is still advisable to protect the delicate parts of ones anatomy from the caustic effects of lime.

The main situation of lime exposure during this process is when mixing the hempcrete. When using a rotary mixer there is a danger of lime dust flying about before the lime is incorporated with the water and this will give an accumulating coating of lime to the skin of your face and any other exposed areas, so it is a worthwhile practice to apply a barrier cream to these areas before you start the days work.

Gloves are advisable when handling the lime sacks especially if you are the one alloted this job for the day. If the worker has any cuts or broken skin, especially on the hands, before or during working, it is very important to apply sticking plasters as lime can be very damaging if it gets into these wounds.

During hot weather nobody wants to wear layers of protective clothing, in this case it is worthwhile protecting the arms from the wrists to the elbows at the very least. Overalls are not usually waterproof and so it is not advisable to wear only this clothing and be naked underneath.

As the hempcrete used in these situations is a fairly dry material the tamping of the hempcrete is quite a clean job, however when washing down the shuttering boards at the end of the day or between lifts there is a danger of wet lime splashing onto the workers and extra care should be taken during this activity.

Protection [Spraying]

When using the spray gun or concrete pump to apply hempcrete materials there is a very great danger that lime will splatter off the wall and enter your eyes. This I can assure you is well worth avoiding!!!!

To best protect your eyes and still see to work, the full face transparent visor is the best choice. It gives the best overall protection and is far easier to keep clean. With smaller saftey glasses it is difficult to fully remove the encrusted lime and they end up more scratched when doing so. A full visor can be dipped in a bucket of water and wiped clean with a soft cloth.

Overalls are a must, as even if you use old work clothes, to keep those you live with happy you will have to remove them before you go anywhere near home as lime dust gets everywhere. The disposable type, though not usually made from biodegradable materials, are in someways preferable as washing lime encrusted overalls will do your washing machine no favours.

When applying hemp plaster manually, extra care must be taken to prevent any of the plaster mixture from getting inside your gloves. Even a small amount of lime trapped between the skin and glove will, where the two rub together, cause a lime burn. This can be happening without being noticed for some time and the resulting burn will leave quite a scar when it eventually heals. This is best avoided by wearing rubberbands around the wrists to secure the overalls over the gloves and prevent wet lime entering the gloves

When moulding or rubbing fresh plaster it is advisable to wear a thin pair of rubber gloves inside a heavier pair.

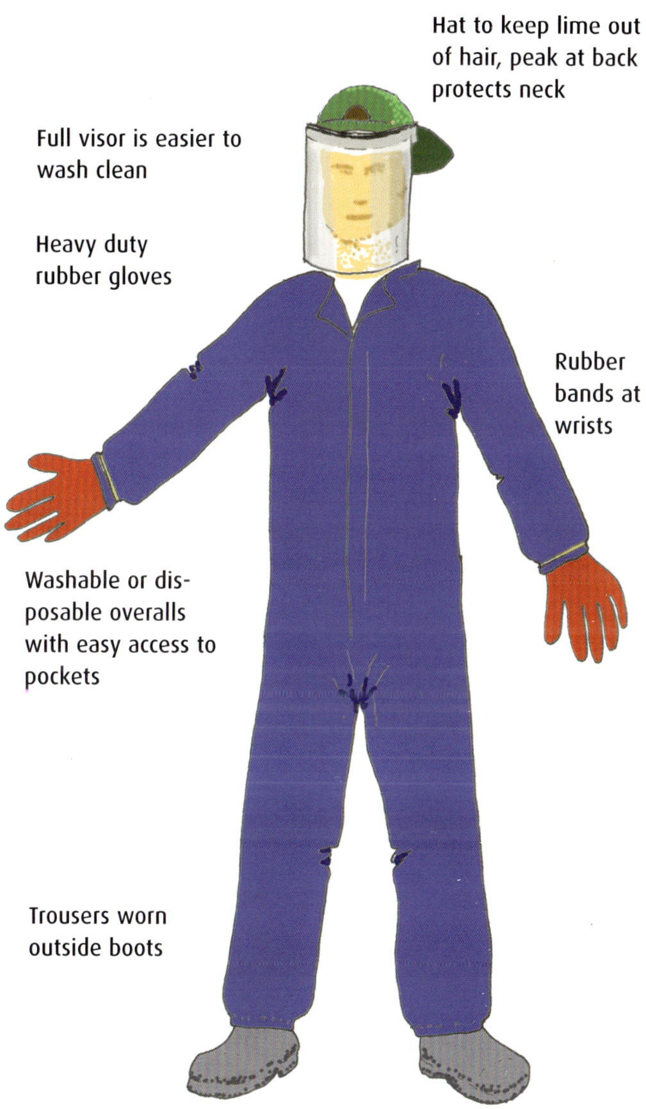

Foundations

As the overall weight of a house constructed with hemp materials might be half of that using concrete or other masonry products, the scale of the foundations needed to support such a building will be smaller. If we consider the ecological implications of the materials to be used several alternatives are open to us. The use of recycled aggregate as a purely dry hardcore base or using stones from the immediate vicinity packed into holes or trenches might be sufficient to provide a base for the further building of limecrete or brick plinthes.

The structural aspect of the foundation other than the support is of course the anchoring of the house. In some cases of wind exposure this will have to be a factor. In such cases anchoring points such as steel straps or bars should be incorporated in the foundation at this stage.

The plinth upon which the timber frame will sit has to both support the structure, but also allow the gravel bed beneath the floor to breathe. This can be achieved in two ways, either by placing vents at regular intervals along its length or by using breathable materials to construct the plinth.

When the ground surface material inside the foundation is soft it should be treated with a dressing of hydrated lime to harden it before being covered with the gravel. If the ground is likely to be wet or damp for most of the year, it will be important to maintain drains around the foundation and to ventilate the gravel bed. The gravel should be of a large enough dimension to prevent any capillary action or obstruction of moving air, but small enough to prevent rodents from travelling through it.

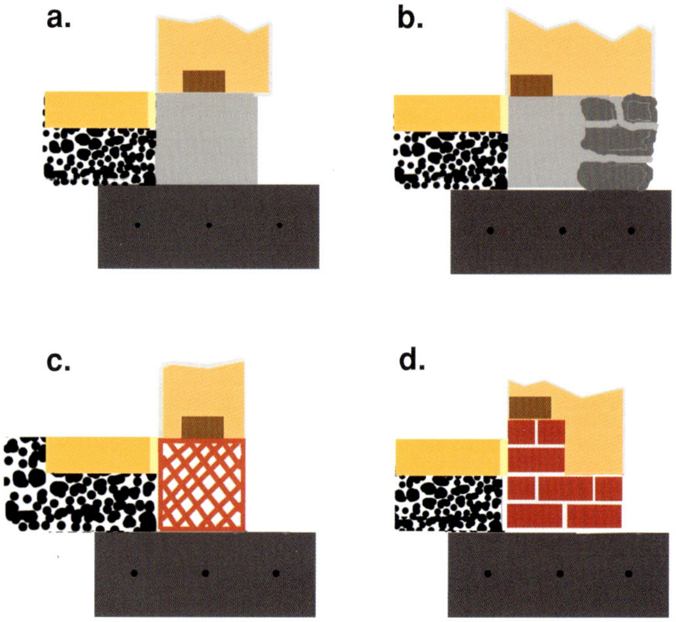

Top; a. Limecrete plinth on concrete base. **b.** Cast limecrete plinth with stone facade on concrete base. **c.** Terracotta block on concrete base. **d.** Stepped brick plinth on concrete base. **Bottom;** Stone wall type plinth using field stones and lime mortar.

Foundations

To cast a conventional concrete strip foundation with steel reinforcing bars, is straight forward. However, if a more ecological route is decided upon then Ground Basic Slag can be added to substitute part of the Portland cement. Alternatively a limecrete can be used. It is advisable to use a powdered hydraulic lime with gravel to produce limecrete as hydrated lime putty will not set properly beneath ground.

The plinth can also be cast with limecrete or created with a row of terracotta blocks, however if these are not locally available stone may be laid using a lime mortar to provide the breathablilty.

It is advisable to prevent damp rising in any way from the foundation up into the timber frame, this can be done by running a plastic d.p.m. along the top of the foundation before erecting the framework or by using a painted bitumen barrier.

Top; Concrete sub-foundation with limecrete plinth.
Bottom; Concrete blocks with bitumen barrier.

Frame Walls

The timber frame of the hemp building serves to give a structure to support the floors and roof and to provide anchoring points for the windows and doors. The hempcrete merely adds ridgidity and fills the frame.

The position of the frame in the wall will depend on what type of frame you are using and what method of application you use to apply the hempcrete:-

1. If the hempcrete is to be sprayed with a pump into the framework and when a permanent facing of strandboard or plasterboard is fixed to the interior surface of the wall, there is no need of battens to be fixed down the sides of the uprights to provide a key for the hempcrete, as the force of the compressed air sticks the material firmly against these surfaces.

2. If the hempcrete is to be applied manually by tamping down and the framework is on the exterior or interior of the wall then a key is provided for the hempcrete to set around.

3. If the framework is set in the middle of the wall then no key is needed as the hemp surrounds the frame.

4. When using poles to build the frame it is advisable to set them in the middle of the wall, as the profile of the poles will be uneven and difficult to use as a surface to fix shuttering onto.

1.

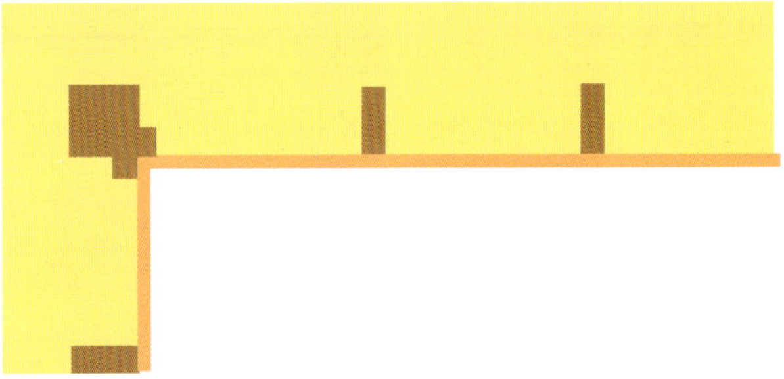

2.

3.

4.

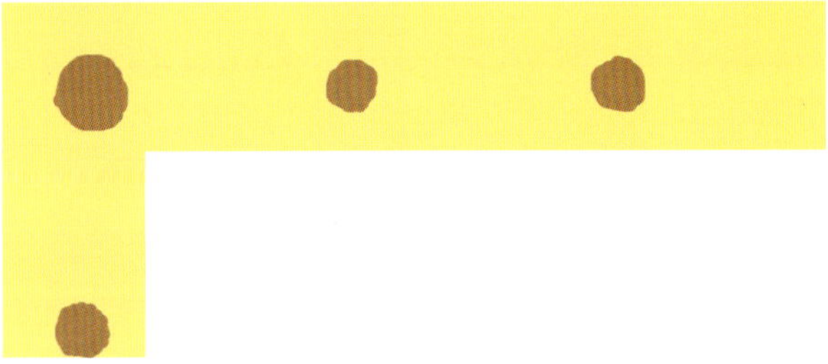

Frame Walls

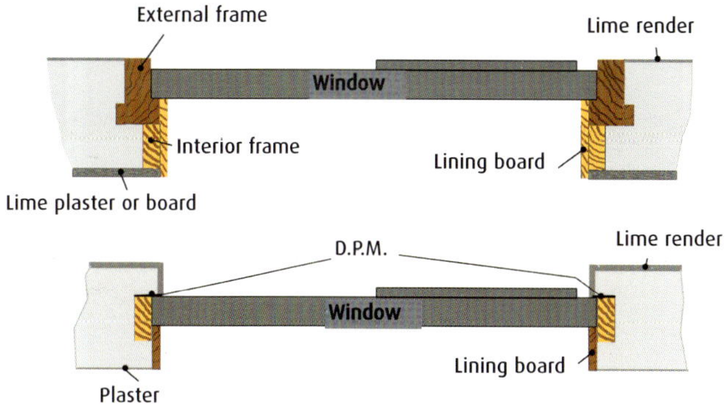

When creating apertures for doors and windows it is essential that they are strong enough to withstand the forces that will be exerted upon them when the doors or windows are in use, especially if a door is slammed shut.

Windows will not only need vertical fixings either side of the frame, but heavy windows will also need to be supported beneath with the frame-work.

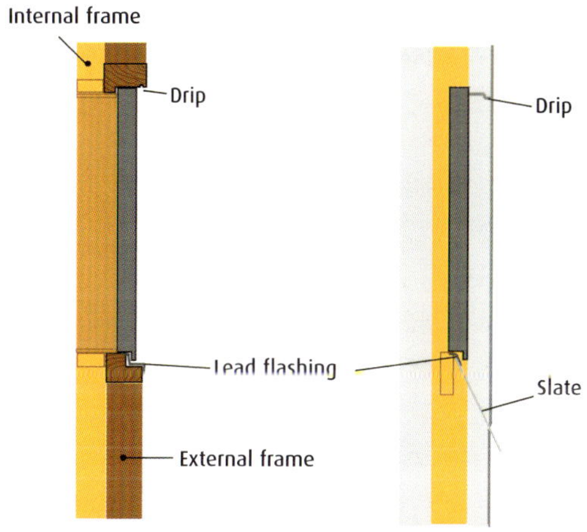

In situations where the window is likely to be subjected to driving rain, the correct detailing will need to be employed to ensure that water is shed around the window and prevented from penetrating.the wall.

Proprietary plastic D.P.M. can be used in vertical strips down the sides of the window within the hempcrete wall. However, at the upper edge of the sill beneath the window, a lead or copper flashing should be used to provide a durable seal at this position.

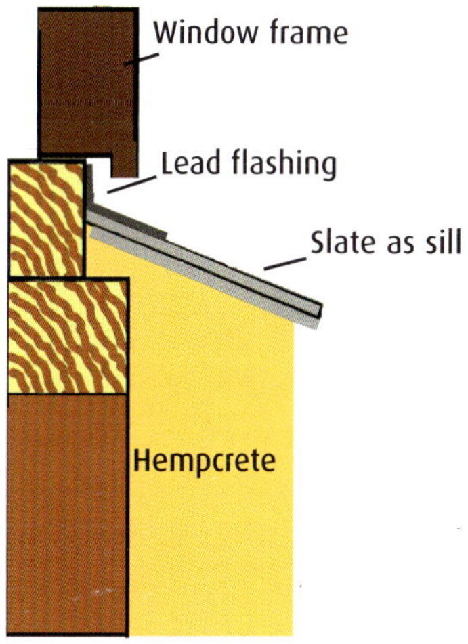

Shuttering

The most important aspects of the shuttering used for building with hemp are; that it should be lightweight and therefore easy to handle and reposition, yet solidly constructed so as not to bow or bend under the forces of compacting the hempcrete against it.

In some cases the project will be a "one off" and the shuttering will only be used the one time. However, when the shuttering is to be used for many jobs, it is worthwhile putting together something a little more durable to be re-used on a regular basis.

For the surface it is advisable to use a material such as marine plywood which is water-resistant and of at least 15 mm thickness. This on its own is not easily handled and might in some situations bow out of line slightly. So, in order to prevent this and to provide a 'handle' it is advisable to fix two timber struts of 50 x 50 cm (as in diagram). This also provides a more solid point to position the fixings.

For the job of securing the shutters to the frame, screws can be used when the shutterboard is to be fixed against the framework, however if the shuttering is to be positioned away from the frame to allow a layer of hempcrete between, it is best to use coachbolts as they are stronger and their hex-heads are more suitable than screws for multiple usage. It is also easier to use a socket type spanner than a screwdriver for the sake of speed.

To take the wear and tear of continued use it is a good idea to reinforce the hole with something such as metal tube. I have found plumbing pipe inserts to be ideal for this job.

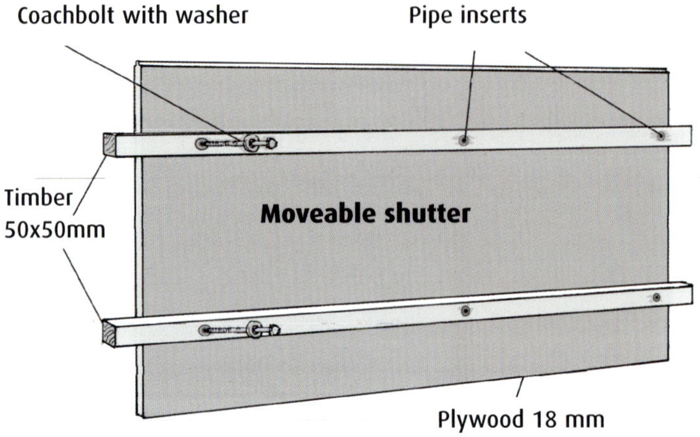

Coachbolt

Pipe inserts

Shuttering

The whole thing can then be painted to make the washing down between lifts easier.

When moving the shuttering boards up for the next section of hempcrete to be cast, the top part of the freshly cast hempcrete wall should be kept compressed by the base of the shutter in its new position so it will not be damaged when applying the subsequent layer.

Sometimes a small section of the top of a wall is difficult to apply the hempcrete from above, due to obstruction by the roof or floor structure. In this case hempcrete is then applied from the side rather than from above against a single shutter board on one side of the frame.

The shuttering system used when spraying hempcrete is different, as there is only a need for a board on one side of the wall.

When applying the hempcrete with a spray gun the element of compression is applied differently and the flow directed so that the hempcrete is not forced between two layers of board causing them to bow outward. However, the impact is great enough so as to make the hempcrete adhere to a surface as smooth as glass, never mind plywood.

To prevent this, a layer of polythene is fixed to the surface of the frame or draped inside the boards before fixing them in place. When a section of wall is completed the boards are removed and the polythene sheeting peeled away from the surface of the hempcrete.

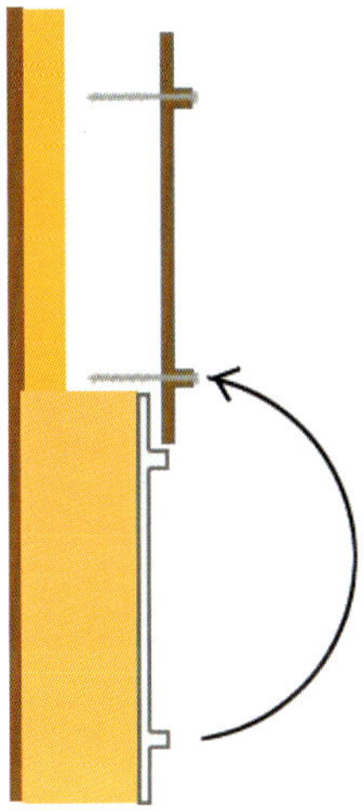

Top left; Positioning the movable shuttering board for the next 'lift' of hempcrete.

Top right; A layer of polythene is fixed to frame prior to shuttering

Left; Once the hempcrete is applied the polythene is peeled away from the surface

The Mixes

FOR A ROTARY DRUM MIXER

All quantities of materials are given in parts per volume.

LIME/BINDER MIX:	
Hydrated Lime	**7 parts**
Hydraulic lime	**1.5 parts**
Cement	**1.5 parts**

LIGHTWEIGHT MIX:	
Hemp	**10 parts**
Lime mixture	**1 part**
Water	**3 parts**

WALL MIX:	
Hemp	**10 parts**
Lime mixture	**2.5 parts**
Water	**4 parts**

FLOOR MIX:	
Hemp	**10 parts**
Lime mixture	**3 parts**
Water	**4 parts**

HEMPCRETE

To mix hempcrete for casting in any of the positions it might be used, a particular blend of limes and binders needs to be formulated to then be mixed with the hemp.

The exact mixture of materials will depend to a certain extent on the climate where they are to be used and the method of application.

In dry warm climates and if the hempcrete is to be applied by hand between shuttering boards, then fast acting binders such as cement or fast acting hydraulic limes should be avoided as large amounts of hempcrete will not be able to be mixed in advance.

In damp, cool climates the quantity of water should be reduced so as to speed up the drying time

When using the concrete pump to spray hempcrete, Portland cements or extremely hydraulic limes should be incorporated into the mix as the hempcrete needs to start to set immediately so that it can keep its position and that the shuttering can be removed very soon after the material has been applied.

When a hempcrete wall is to be left exposed without a covering of plaster or board, it is more pleasant to be looking at a lighter creamy colour rather than the grey that is produced by Portland cement, in this case white cement can be used.

In situations where the thermal capacity is to be increased, such as an internal wall, the quantity of sand should be limited to 20% of the volume of the mixture.

HEMP PLASTER

To mix hemp with a lime based binder in order to use it as a plaster it is preferable, if applying the substance by hand, to use a hemp chip that has been sieved so that only the smaller particles (ie. 1-5 mm in length) remain.

If using a spray gun to apply the hemp plaster, then the normal particle size (5-30 mm) can be used if a particularly smooth surface is not required.

The plaster mix has a higher quantity of hydrated lime in it, in order to give a more plastic consistency to the material so that it can be spread out on the wall more easily.

When using hemp plaster in hot dry climates, ground limestone can be added to the mixture to retain moisture for longer.

When applying hemp plaster to a hard background, such as concrete or dressed stone, the wall should be sprayed or rough plastered with a coat of lime render containing a hydraulic lime with a high compression strength first. The NHL contained in the lime mix should also be of this same high value, this is so that the hemp plaster adheres properly to the background.

To make the plaster harder and more able to take knocks the addition of sand is recommended.

The application of hemp plaster with a spray gun is only possible with the addition of sand. This is to give the material sufficient weight to enable the plaster to fall down in the hopper and into the chamber to be propelled by the compressed air.

FOR A ROTARY DRUM MIXER

All quantities of materials are given in parts per volume.

HEMP PLASTER LIME/BINDER MIX:

Hydrated lime	**8.5 parts**
N.H.L.3.5 or 5	**1.5 parts**

PLASTER MIX (MANUAL):

Hemp	**10 parts**
Lime mixture	**9 parts**
Water	**5- 6 parts**

PLASTER MIX (SPRAYED):

Hemp	**10 parts**
Lime mixture	**7 parts**
Ground limestone	**2 parts**
Sand	**2 parts**
Water	**5 parts**

Mixing

To mix hempcrete in a wet state it is possible to use a common drum mixer or spiral trough mixer.

To ensure the correct and continuous mixing of hempcrete using a rotary drum mixer the following steps must be followed:

1. Pour the required water needed for the type of hempcrete you require into the stationary drum.

2. Pour the measured amount of lime into the drum and start the drum rotating. It causes less dust if the drum is stationary when adding the lime. Allow a few minutes for the limes to mix together and form a paste with the water.

3. Add the hemp gradually and leave to mix until it is completely coated with the lime. This should take 3-4 minutes after the hemp has been added.

4. Tip the hempcrete out of the mixer into a wheelbarrow, bin or when mixing continuously onto a board laid on the ground. For the consecutive mixes, add the water and then turn the mixer on and allow the water to wash the remnants of hempcrete off the walls of the drum before stopping the mixer to add the lime.

When using a trough or pan mixer the hemp and lime can be added and partly combined first and then the water added.

Tamping

To ensure a uniform consistency to the hempcrete in the wall, both from a lime-hemp ratio and structural point of view, it is essential to get a system going.

If the quantities of lime or water in the hempcrete vary too much this will be visible in the wall when dry, and it will of course not behave in a uniform way. Apart from controlling the quantites being added to the mixture, making a large amount of hempcrete before you start to fill the shuttering and mixing it around as you take shovelfuls from the pile will help.

To ensure the hempcrete is compacted evenly it is necessary again to utilise a system of application. If the hempcrete is tipped into the shuttered cavity in an uneven manner and then tamped before levelling out, then the peaks of the material will be compressed and the lower points will not. This will be repeated as the wall continues to rise causing pockets of untamped material which will fall apart once the shutter is removed.

Positions around plumbing pipes and electrical conduits must be given special attention when tamping to create a strong hempcrete mass around them.

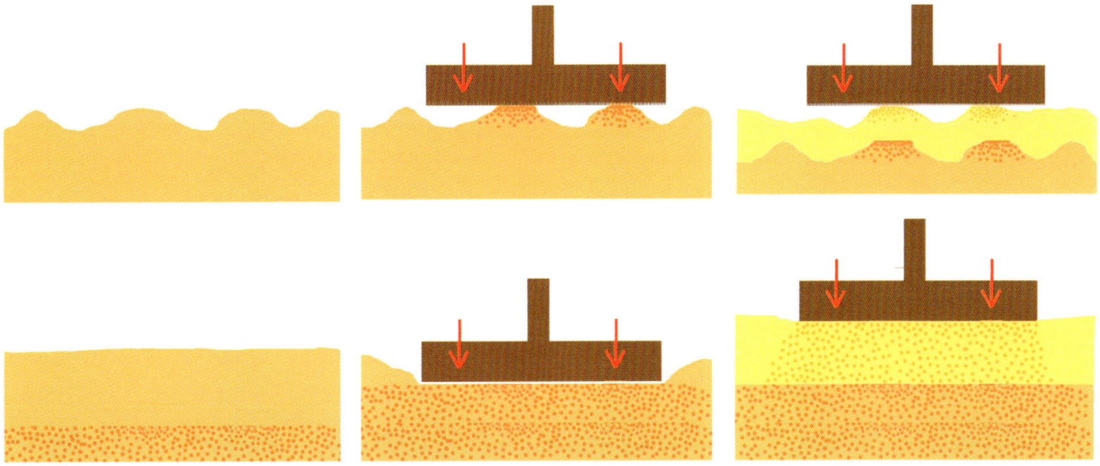

Ground Floors

When choosing the hemp system of floor structure the condition of the ground, within the foundations, upon which the floor is to be layed must be considered first.

If loose and prone to burrowing by rodents, broken glass and lime powder can be incorporated into the top layer of soil and tamped down hard before the gravel is laid.

The gravel should be clean and of between 20-100 mm in diameter. This is to ensure that when it is packed down there is enough air-space to ensure no capillary movement of water can occur up through the gravel-bed and that there is a small amount of air movement.

The floor hempcrete mix is made with only sufficient water to be able to bind the material together, so no 'fat' or lime rich layer will form on the surface as happens with concrete floors.

Before laying the hempcrete out on the gravel, guides or markers should be set up to help keep the hemp layer level. It doesn't need to be perfect as when laying the limecrete screed or mortar layer at the next stage will accommodate any slight irregularities, but it does need to have no lumps or dips in it.

When laying a hemp floor beneath underfloor heating pipes during the winter months the drying out can be assisted with a gradual application of heat, if the heating source can be utilised.

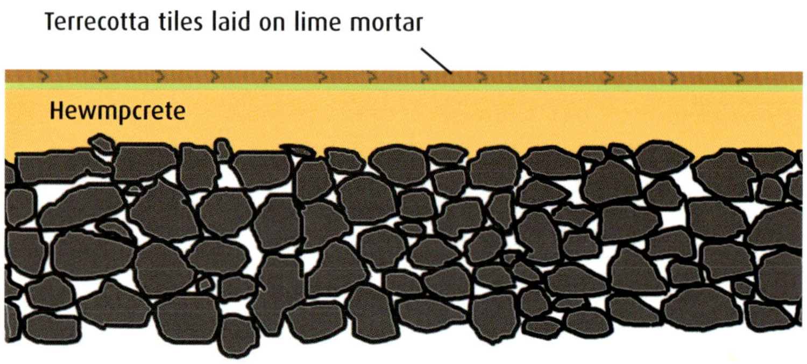

Above; Hemp floor mix on gravel.

Floor Wall Tiling

Tiles used for walls and floors need to have entirely different characteristics when used in a hemp building. Those on the floor have to breathe. While those on the wall, where they will be subjected to large amounts of water especially around baths and showers, generally will not.

When laying terracotta floor tiles it is advisable to first give the tiles a coat of linseed oil, applied with a cloth, to prevent lime staining the tiles when working.

The mortar used to cement the tiles can be made quite dry from a mix of; 1 part hydraulic lime to 2-3 parts sand.

This is applied in a layer of at least 20 cm. At this thickness any uneveness on the hempcrete or limecrete layer beneath can be rectified.

The grouting can be done as you go, so as to avoid having to clean up twice. A grout can be made from a mixture of either pure hydraulic lime with enough water to make a stiff paste or if the gaps between the tiles are over 4 mm a mix of 1 part hydraulic lime to 1 part sieved sand. Choose a sand of an appropriate colour so when mixed with the lime it produces a complementary shade.

When applying wall tiles with a hydraulic lime paste apply enough for 2-3 tiles at a time. Press firmly into place once the tile is in position. Do not try to remove and re-apply. You might find you have to dampen both the tiles and the wall before fixing tiles.

Use the same paste as a grout.

Top; Lime spread with toothed trowel.
Middle; Tiles should be positioned firmly in one go.
Bottom; Terracotta tiles on deep bed of mortar.

Between Floors

In most cases the cavity between floor joists is best filled with the lightweight hempcrete insulation mix.

All first fix installation of electrical and plumbing supplies should be installed previously.

The hempcrete should be lightly tamped down around any electricity cables or plumbing pipes to a level slightly above the top of the joists. When the flooring boards are fixed down, the mix will then be compressed so as to leave no cavity which might cause unwanted echoing.

If the beams are of special visual value and it is required to leave part of them exposed, a beading can be run along the sides of beams to rest panelling upon to support the hempcrete.

These short lengths of panelling do not need to be fixed as they are kept in place by the hempcrete above.

The shallow layer of hempcrete might reduce insulation levels, both for heat and sound compared to completely filling the cavity from the bottom of the floor joists to the top, however it is usually enough to give sufficient levels of both.

Top; Strips of beading running along the inside of joists to support boards.
Right; Beneath a hempcrete filled floor.

The technique of spraying the hempcrete can also be utilised in this situation.

Care should be taken that any components are firmly anchored down as the force of the compressed air could blow them about the work space.

It is the practice when building using lime to lay fibreboard down on the rafters first to take the wear and tear of the messy work of plastering, and the inevitable dirt trodden in by the personnel on the site. The finish floor surface is layed when all this has been completed.

Partition Walls

The lightweight mix of hempcrete is also used for internal walls. In this case a stud frame is erected and the plasterboard fixed to one side. In order to be able to reach down inside, when tamping the hempcrete, smaller plasterboard batts (or larger sheets cut down) are then fixed to the side you are working from and form the shuttering to pack the hempcrete between.

This is done with only sufficient force to fill the cavity or the plasterboard will bow out.

Alternatively, the lightweight hempcrete can be sprayed between the frame and against the board fixed to one side of the frame. The hempcrete is leveled to the exposed surface of the frame and then faced with more board.

Top & Left; Lightweight hempcrete mix infilled between joists.

Roofs

The structure of the roof must be suitable for the type of frame supporting it. Poles or sawn timber can be used so long as a sufficient depth of insulation can be accommodated.

If hempcrete is to be utilised, the elements of the roof structure will have to be strong enough to support slightly more weight than would be normally calculated for, using traditional insulation materials.

Hempcrete can also add a small element of structure itself, if the hemp and lime mix is well tamped around the timber frame. Or, if the wall mix is used, it can be cast as a dome or supporting arch and be self-supporting when the mould is removed.

Take care to provide footholds if the hempcrete is to be placed manually by workers on the roof.

When applying hempcrete with a pump method it is most efficient to level the hemp material to the top of the rafters as it is already compressed with the force of the compressed air, however if the material is being installed by hand it can be tamped down to the required level beneath the breather membrane.

It is always better to keep insulation materials away from the underside of a breathable membrane as there is the risk of condensation forming before the moisture has evaporated.

Right; Lightweight hempcrete infill applied manually leaving air space beneath breather membrane.

Right; sprayed hempcrete supported on rough sawn boards which are left exposed until drying is complete when they are then covered with finished boards.

Roofs

When applying hempcrete in a roof cavity it is far easier to work from above when there are only rafters and boards beneath, though care should be taken to ensure that workers do not have to step between the rafters.

In this situation access is gained to the whole roof cavity and the hempcrete can be installed quickly without having to bring the material into the building. This is true whether the hempcrete is being sprayed or mixed damp and shovelled into place. Hemp fibre insulation panels could also be installed from above if the weather permits.

Once the hempcrete has been installed all windows and apertures should be opened to allow the maximum movement of air to dry out the materials.

When installing hemp fibre insulation batts take care to ensure that the spacing of the rafters is slightly narrower than the width of the batts, as they can then be easliy wedged between them and will not need to be further secured with staples or other fixings. Remember masks should be worn, especially when working frequently beneath such roof spaces. The small fibres that will fall off the insulation material are not harmful in small quantities but workers in the scutching pits of the mills a century ago suffered from lintosis caused by prolonged exposure to breathing in the fine strands.

Above; leveling hempcrete before tamping down and the fixing of counter-battens
Below left; section of engineered timber "I" Beam

Below; Diagram of roof components: Tiles on battens on counter-battens on breather membrane on counter battens on 'I' beams infilled with lightweight hempcrete. Plasterboards or other fixed to soffit of 'I' beam rafters.

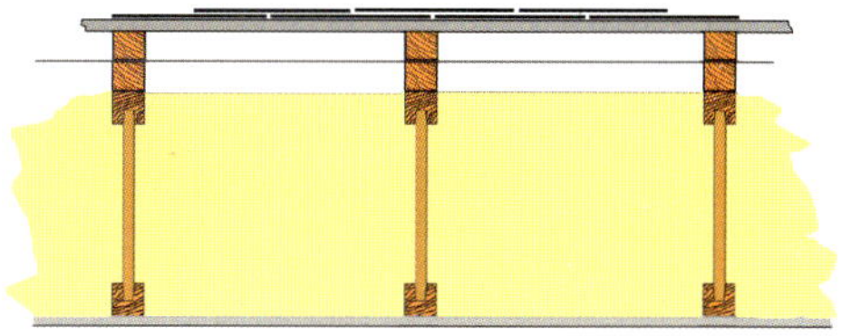

Spraying Hempcrete

In recent years a method of applying hempcrete by spraying against formwork has been developed by a company in Brittany named C.T.C.
This reduces the time taken for this element of the job considerably.

Above; Mixer and pumping equipment.
Below; Water controls on pipe

During the operation of this combination of machinery the hemp and lime are placed in the hopper which then is lifted hydraulically and tipped into the adapted mixer which, in a matter of 3 minutes or so, has completed the combination of the materials and then steadily empties the dry mix into the hopper on the top of the pump. By means of rotating chambers, the dry hempcrete is then blown along the pipe with considerable force.

Water quantities are controlled by the operator by means of valves at the end of the pipe.

This method has several advantages over the manual method other than speed, in that the mixture requires less lime and water in the mix which increases the insulation value and lessens the drying time.

Spraying Hempcrete

Other differences to this system compared to normal shuttering with hempcrete, are that due to the direction of the force, the hempcrete is being propelled into the framework and against the shuttering boards. These boards can be of lighter dimensions. With the tamping method, the material is pressed downwards, which bonds the layers of material together. However it can, if overdone, force the shuttering boards either side apart causing a bulge in the wall.

With the spray method this doesn't happen, although the impact of the air driven material is such that it is necessary to drape a sheet of polythene inside the shutterboard to prevent the hempcrete adhering to too great an extent. This is then peeled off the surface of the hempcrete when the shuttering is removed

Once the hempcrete has built up to the required thickness against the external or internal boards (depending on which way round the job is being done) the surface of the hemp is then raked flat and levelled off with a straight edge.

Top; Hempcrete being sprayed by Laurent Goudet

Left; Levelling off

Far left; Surface of sprayed hempcrete

Spraying Hempcrete

Before this work of spraying is carried out, all woodwork, doors or window and surrounds together with lintels or rafters whether painted or not, should be masked with polythene or tape. As to clean the hemp and lime off afterwards is very labourious.

A first coat of lime and sand plaster can be sprayed onto a hempcrete surface immediately after it has been applied. Hempcrete applied internally during renovation, or a complete wall of the material, is best protected by such a coat of plaster as it helps the hempcrete to dry out.

The most efficient way of doing this is to spray it on using a compressed air gun made for such a purpose. As it is not essntial to completely seal the surface of the hempcrete, the plaster can be sprayed quite quickly. The mix can be quite dry so as to give a rough surface to provide a key for the next coat

Later when the wall has dried out the final coats of lime plaster both external and internal can be applied. It is better to wait till this point, as during the drying process tannins might leach out and discolour the plaster.

Left; Timber lintel and window covered with plastic sheeting for protection whilst spraying.

Below; Sprayed hempcrete with first coat of sprayed lime plaster

Spraying Hemp Plaster

Protective clothing and equipment must be worn when using the spray gun! There is especially a risk of lime entering the eyes and I can assure you this is not recommended.

A plaster spray gun can be used to apply hemp plaster in the same way as it is used to apply a conventional plaster of lime and sand.

The gun itself consists of a hopper and chamber with nozzles for the compressed air to enter the mixture. It comes supplied with a number of plates with varying sizes of holes for the plaster to leave the gun. When used for hemp we require the plates with the largest of these to allow the particles of hemp to be blown out.

The mix is also altered by the addition of sand. This lowers the insulation value slightly but is necessary to provide weight to the material so that it falls down inside the hopper as it is being blown out.

The addition of sand also produces a more gritty surface, though it is harder and flatter. It is worked with a wooden or plastic float to the required finish.

Top; Spray gun used for spraying hemp or lime plaster.

Left; Sprayed hemp plaster finished flush with timber reveal detail.

Hemp plaster by hand

The most important aspect of plastering with a hemp plaster is the consistency . This does not depend on a one-mix does all approach, it will depend on what finish is desired, the type of background to be plastered and the atmospheric conditions.

Although the constituents of the mixture should be well combined, it is not advisable to leave the plaster mixing too long in the machine even though it might seem as if it is a creamier consistency. Too much air in the mix will cause the plaster to set as a frail sponge-like structure and not be strong enough.

The surfaces to be plastered must be assessed for the quantity of water they will soak up whilst plastering, as hemp plaster will not adhere to too dry a surface. Test an area first and if the wall is very absorbent, wet a small area thoroughly before plastering it and then go on to the next part.

If the wall is of moderate absorbancy then larger areas can be sprayed with water and plastered in longer sessions.

The method used to put the plaster on the wall will be a personal preference. I find it usually depends on what, if any, plastering experience has been had previously. Professional plasterers are used to a very malleable material they can move around their hawks and apply with a float like butter, hemp plaster is not like this.

The finer grades of hemp, seperated by the processor for plastering are much easier to use and a finer finish is more easily achieved.

Hemp plaster by hand

1. Wall with loose rubble filling up to new roof.

2. Hempcrete wall mix applied to fill gaps around loose rubble and to level surface of wall.

3. Hempcrete ready for hemp plaster

4. Hemp plaster daubed onto wall.

5. Surface of wall is smoothed and hollows filled in.

6. Surface of hemp plaster worked with float to create level surface and to bring fat of lime to the fore.

7. Hemp plaster is left to start to set for a night.

8. Next day the surface can be rubbed to embed the fibres in the lime and create a smoother surface.

External Pointing & Renders

Buildings that have hemp walls or old stone houses plastered with hemp internally should be finished externally in a manner that is appropriate.

Stone buildings should be pointed or rendered so as to pervent water penetrating the wall but also to allow it to breathe.

Pointing a stone wall with lime mortar should be done in a way suitable for the type of stone used in the area. It is worthwhile studying the traditional local styles.

Hempcrete walls should be rendered with at least 2 coats of lime render, the second somewhat weaker, that is with less lime in the mix or using a lime that has a lower compression strength than the first coat.

Suitable ratios of sand to lime are:
1st coat; 1 Hydraulic lime-4 sand.
2nd coat 1Hydraulic lime- 5 sand
Hydrated lime can be mixed with the hydraulic lime on the final coat, but as it has more free lime is likely to leach out at the surface somewhat and so is only suitable with surfaces that are to be painted white.

The first coat is keyed to give purchase for the second. The manner in which the final coat is worked will depend on the appearence required. It can be smoothed with a wooden, plastic or steel float or, if sprayed or dashed left as it is.

Left; Lime mortar used to point stone wall
Middle; First coat of lime render applied to hempcrete wall to be scratched.
Right; Second coat of lime render on scratch coat.

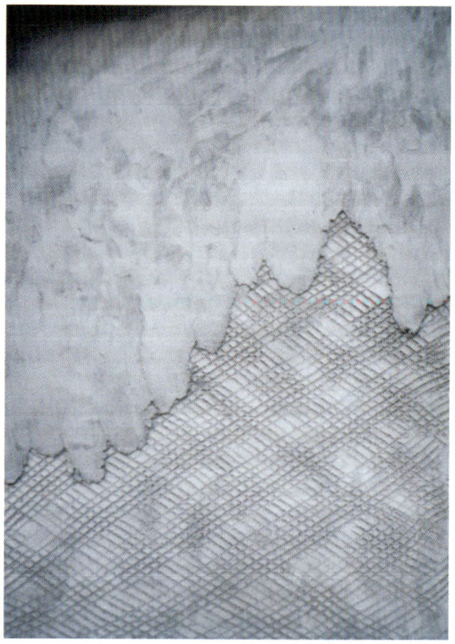

Hanging Slates & Tiles

To hang tiles on the exterior of a hemp wall the framework should be positioned on the outside face. It is then possible to mount the battens on the exposed surface of the uprights.

A variety of means by which the tiles or shingles are fixed to the battens can be used. Specialised clips can be purchased that are fixed to the battens, the tiles are then fitted into these. In addition the slates can be nailed or screwed in place.

If they are to be nailed it is better that the hempcrete part of the wall has completed its initial setting before being subjected to this amount of vibration. This is not necessary if they are screwed in place.

Unless this system of wall cladding is used in the most exposed of situations, it is unnecessary to use a breather membrane beneath the slates or tiles

The detailing around doors and windows and beneath eaves and verges must be sufficiently watertight to prevent continued penetration of rainwater.

Top; Hempcrete wall with external frame.

Middle; Diagram of battens fixed to frame.

Bottom; Detailing around window.

Exterior Painting

When painting the outside of a hemp building the usual rules apply concerning the breathability of the paints. Even with the timber components it is better that all the structure can breathe.

With lime paints the first coat is applied in a diluted form, using brush strokes in every direction to ensure the paint covers all hollows, the second coat can be applied straight after. This is finished with vertical strokes to assist the shedding of rainwater.

Alternatively, other natural paints can be used to coat the timbers. These may be oil-based, lacquers or stains, so long as they are suitable for external use.

If you require a more maintenance free treatment for the rendered sections of the exterior, then what are termed pure mineral paints can be employed. These soak into the surface of the porous render and need to be used within two weeks of being mixed. They cannot be removed, only painted over, and last for years in the same way as pigment added to the render does, by the fact that the fresh colour is constantly being exposed as the wall weathers.

When using eco-friendly paints always read the manufacturers instructions as regards safety, as although natural or traditional paints are less toxic to the environment, they are nonetheless unpleasant when they enter the eyes or broken skin.

Left top; Lime paint applied over diluted first coat

Middle; Second coat finished with vertical strokes.

Bottom; Old houses painted with lime paints.

Oiling Timberwork

If the timber sections of your building are to be exposed as a feature, then it may be preferable to leave them unpainted. To both protect the timber and enhance the appearence, it is advisable to apply an oil.

For both exteror and interior application boiled linseed oil can be used. However in positions where the surface of the timber will need to stand up to a great amount of wear and tear, it is probably better to use a product such as tung oil, internally.

Great care should be taken when applying the oil to the timber. It is advisable to prevent any oil from getting on to the hemp sections of the wall where it soaks in very quickly into the hempcrete and it is then difficult to remove the stain.

This can be achieved by applying the oil using a small brush and masking the areas you don't want the oil to be applied to with masking tape.

As oil repells water, it is preferable to paint the sections of hempcrete wall with a water-based paint after the timber has been oiled, to save labour cleaning off the paint. The oiled timber can be masked with tape before the painting.

Above; Oak framed medieval house with timber treated with oil.

Left; Internal framework oiled to enhance appearance.

Electrical Installation

All installation of electrical circuits should be carried out by a qualified person, however the components, such as conduits and plug and socket boxes, together with features of the framework such as the holes for the conduit to run through the upright timber elements, can be installed as part of the frame building.

It is advisable to use flexible conduit for all electric cable runs. This can be cut to the required length and run through holes made in the framework immediately the frame is in place.

Remember to thread a string or cord through the conduit first to enable the electric cable to be pulled through later.

If the weather conditions permit, it is possible to measure the cable lengths at the same time as cutting the conduit and threading the cable through, before fitting in the framework. This would, in the case of building multiple units, very much reduce the time spent on this part of the job.

All tail ends and boxes should be protected from the ingress of hempcrete whilst it is being applied.

All second fixing is done is the same manner as other building techniques.

It is advisable to wait until the hempcrete has dried to turn the power onto curcuits.

Above; First fix electricity supply, installed in the framework prior to casting the hempcrete.

Light Fixtures

The best way to bring out the visual qualities of a hemp wall is of course, with light. Mounting a shade to deflect the light across the surface will highlight the texture.

The use of ceramic fixtures works well on a hemp plastered wall. A basic light fixture is mounted at the surface and the shade simply screwed to the wall surrounding it.

When moulding a shape from hempcrete it is best to use a mix with slightly less lime than the plaster mix in order to maintain some translucency. The shape of the light shade is slowly built up and out from the wall, in much the same way as a swallow builds its nest. Small amounts of hempcrete are applied in courses using the fingertips and moulded into the required shape. Further smoothing is done when the hempcrete is setting.

Some wall mounted, ceramic light shades are translucent and can be mounted on the wall during the plastering work. The hemp plaster can then be brought up around the edges to make the fitting seem to be emerging out of the hemp plaster.

Top; Ceramic light shade mounted on the surface of hemp plastered wall.

Right; Diagram of cross section of moulded hempcrete light shade.

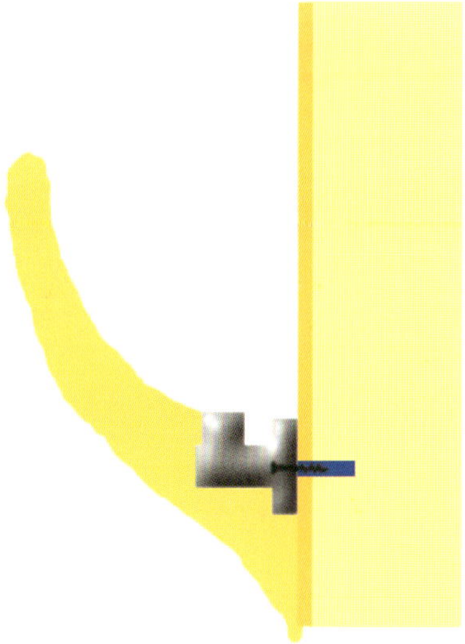

Underfloor Heating

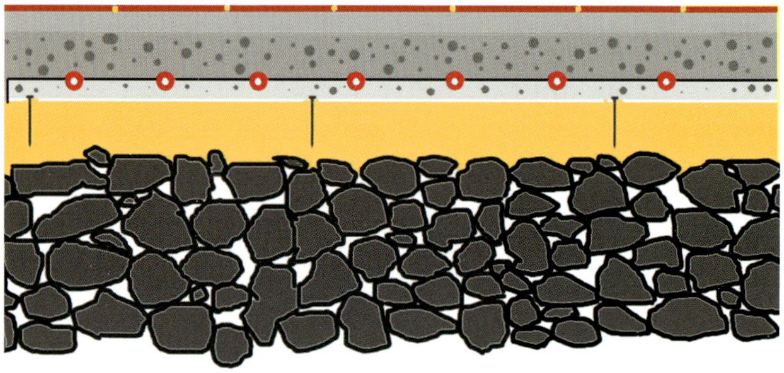

Above; Diagram of hemp floor with underfloor heating pipes in limecrete screed on hempcrete layer.

Below; Pipes fixed to hempcrete floor.

The use of underfloor heating has become very popular in recent times. With the possibility of avoiding the use of radiators, thus freeing up wall space and the fact that many radiators are not really seen as an attractive addition to the look of an interior, underfloor heating is often now the choice of architects and interior designers. Radiators are also thought to create unhealthy circulation of air causing problems for those with dust allergies. There are also those who claim it is not healthy to have ones feet warmer than the rest of the body. Despite these conflicting ideas many people like the idea of having a warm floor to walk on.

This can be achieved with the hemp building system by fixing the pipework to the surface of the hempcrete layer in the floor using the standard strips, manufactured for this purpose, that the pipes clip into. These are fixed with screws or nails driven into the hempcrete.

The pipes are then covered with a layer of a minimum of 100 cm of a limecrete screed to produce the thermal mass needed for the heating system to work correctly.

Tiles are the best choice as a covering for such a floor, however timber can be used provided the floor is completely dry before the timber is laid down upon the screed. The heating system within the floor is very good for accomplishing this drying action, so long as the heat is applied slowly at first so as not to crack the limecrete.

Radiators

A radiator works very effectively with a hempcrete wall. To fix the radiator securely to the wall it is necessary to provide a fixing point within the timber framework. This will mean either positioning the uprights in line with brackets of the radiator, if the size of radiator is known beforehand, or fixing a horizontal piece of timber at approximately the position in the wall that the radiator is likely to be hung.

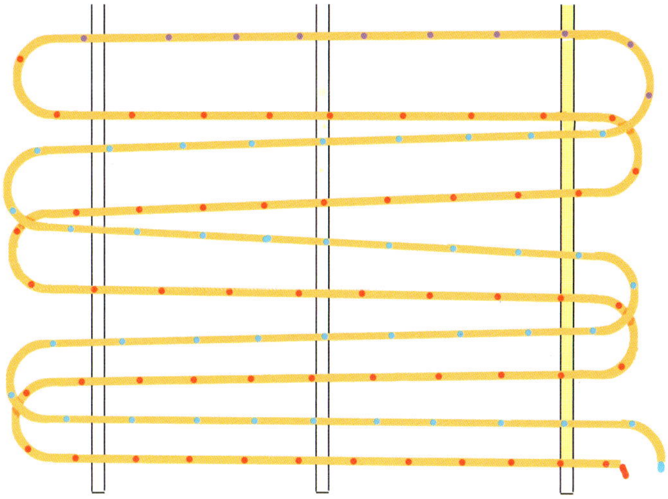

Another option is to use underfloor heating pipes set into the plaster as mentioned before in the previous section on the renovation of castles, this is a choice that is perfectly suited to situations where the installation of underfloor heating is impossible and the use of visible radiators unsuitable.

Top; Diagram of heating pipes mounted on wall.

Right; Cross section of fixing of pipes to wall.

Below; Pipes in place on hempcrete wall.

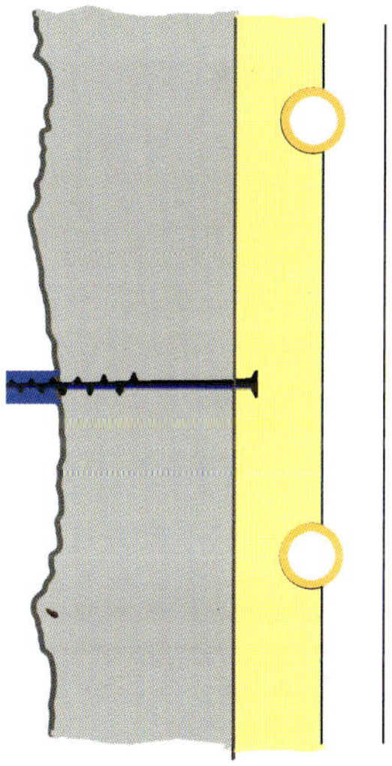

The walls should first have an insulating mixture of hempcrete applied and the channels for fixing the pipework will have to be securely fixed through this into the wall.

The heating pipes are then clipped into the channels ensuring that the flow of water will work without creating airlocks.

These pipes are then covered with a lime-sand plaster which needs to be applied in layers to achieve the sufficient depth to cover them completely.

Plumbing

The installation of plumbing services in a hemp building are no more complicated than in any other type of building and in many instances are a lot easier.

As long as all the positions of baths, showers, sinks, toilets and washing machines are planned out to start with (and those plans adhered to) the pipe work can be installed in the framework before the application of the hempcrete.

From a structural point of view the pipes should be surrounded by a minimum of 50 cm of hempcrete.

All polyethylene and polybutyle pipes have been tested for resistance to alkaline substances and are suitable for use with the hemp building system.

Copper pipes can also be incorporated in a hemp wall and some would claim that these are more ecologically suitable as they can be more easily recycled at the end of their lifetime.

All joints should be fully tightened and if possible tested before the application of the hempcrete as leaks will cause just as many problems in a hemp building as they will in any other type of building.

Any open ends to pipes should be sealed to prevent the ingress of the hempcrete during application.

Smaller pipes of 30 cm or less should be fixed to the framework with proprietary clips. Larger pipework such as downpipes and those for waste outlets can be affixed to the timber frame with

Top; Waste pipe fitted flush with surface of wall during casting, to be joined to drain later.

Below; Waste water downpipes fixed in framework prior to hempcrete application.

Plumbing

clips, straps or if desired twine (for a more ecological solution), as the hempcrete, when set in place, will hold them in position.

If the position of a bath, sink, toilet or whatever, is changed after the hempcrete has been applied, the advantage of hempcrete is that it can be easily excavated and the pipework altered. More hempcrete can then be cast around the altered pipes. This also applies to new additions to the plumbing system.

Plumbers have always used hemp,(however not in the form of hurds with lime), as they have for a long time employed the long strong fibres to seal the joints of the pipework, sometimes with the addition of a paste, by wrapping the fibres around the threaded compression joints to completely seal the joint when tightened.

Above; Hank of hemp fibre used by plumbers to seal pipe joints.

Below left; Outlet from toilet is easily made through hempcrete wall.

Below; Pipe lagging made from hemp fibre.

Do's & Don'ts

As with all building systems and materials there are limits to their use, and rules to obey if there are to be no problems with using them .

The hemp building system is just the same and though much of it is common sense there are a few points that need to be made.

Some people assume that hempcrete can be used as a cure-all remedy for problems with an existing building or in any position on a newly built house, unfortunately this is not true. As with all structures made using natural materials, moisture can have detrimental effects and this is just the same with hemp.

Throughout this book I have made references to the need for hempcrete, and those materials used in conjunction with it, to breathe. This is very important and the reason why I have repeated it so often.

As with timber, hemp hurds if they are kept at 20% moisture levels or above, will be attacked by wet rot or other fungal infestations. When the hemp is growing in the field, the spores of many species of fungi will adhere to the stems and will still be present after the materials have been separated into fibres and hurds.

If the drying period of the hempcrete is prolonged and the spores are present, the fine thread like structure of the mycelium will start to spread through the

Above left; Face of hempcrete wall left exposed to the elements for 6 months showing free limes leached out of the material to form layer on surface.

Left; Ischnoderma fungi growing at edge of hemp field, the spores of which are often found in hemp hurds

hempcrete. This can be observed on the surface of the hempcrete by the appearence of various discolourations depending on which species of mushroom we are dealing with.

I have witnessed many different types of mushrooms sprouting from raw hurds spilt on the ground during a building project, some of them edible such as the ink cap, others not, but those that are likely to cause the greatest problems are the same species that attack wood in buildings.

If these fungi start to grow it is only because the hempcrete still contains too much moisture and so needs to be encouraged to dry out more quickly. Alternatively, there is a problem with moisture entering the structure that needs to be addressed .

In ideal circumstances the hempcrete should be brought down to 18 % moisture within 6-8 weeks. If this is not the case, heat and moving air will have to be introduced around the structure to reduce the moisture levels. For this reason it is advisable to start a building project when the weather is suitable to do so.

Other causes of problems associated with moisture are that the hemp component of the mixture is contaminated with too great a percentage of leaf material. This can cause areas of rot to occur as the leaf particles more quickly decompose in the presence of moisture.

Other problems occur when, during packaging, the hurds become compressed together in clumps and are then not broken up before being fed into the mixer. These are then not sufficiently coated in lime and do not have the mineralising action of lime.

Do's & Don'ts

During the building process water can cause other effects which are not necessarily problematic, such as hempcrete walls exposed to rainfall over prolonged peroids can cause the free limes to leach out and form a skin on the surface of the wall. This will be covered with a lime render or other exterior cladding and cause no problems once the wall has dried out. Exposure to rain might also cause the tannins in the hemp to migrate to the surface causing a pale brown discolouration. This can cause further staining of subsequent layers of plaster or paint, but can be prevented from doing so with an extra coat of hemp or lime plaster when the original area has fully dried out.

The timber used in the framework should be of a type that is not likely to warp or shrink while the hempcrete dries out. If, for some reason a part of the frame does suffer from this problem the resulting gaps can be infilled easily with more hempcrete, so long as the structure is not compromised. In any case it is best to leave the final coat of plaster or render until any likely movement of the framework has stopped.

When an area of hempcrete wall or hemp-plastered wall will eventually be covered with a very large picture or mirror, such as the dance room pictured below, it will be necessary to ventilate the space behind it, if the climate produces high levels of humidity.

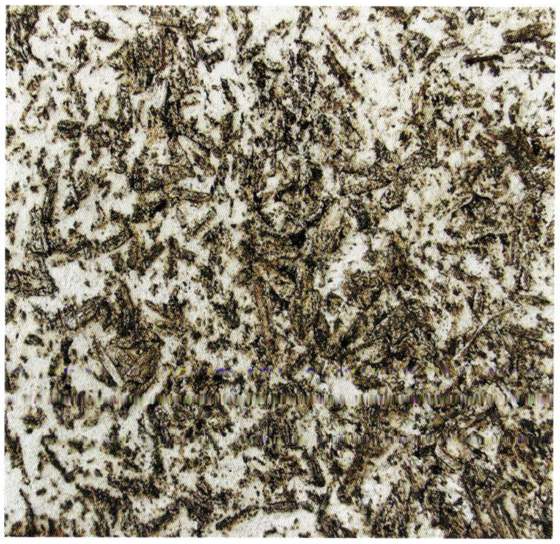

Above; Hempcrete stained with tannin.

Right; Hemp plaster kept wet for too long causing hemp at the surface to grow moulds.

Below; Large mirror on dance room wall.

The Wendy House Project

When my daughters asked me to build them a play house, it occured to me that this was an ideal opportunity to build a simple structure using hemp.

So, starting with a basic frame of birch poles, I drove the upright elements of the frame into the ground and using twine, tied more poles across these to form a wall plate. I then formed window openings and a doorway. On the top of this I tied more poles to form the roof structure.

To keep the bottom of the hemp wall off the ground, I laid stones in a line between the poles and filled the floor area inside this with gravel to provide a base for the floor.

Using the hemp wall mix, the walls were moulded by hand from hempcrete, pressing it firmly into place as there was no shuttering to force the hempcrete between.

Before covering the roof, twine was attched to the roof structure to later tie the thatch down onto the frame. When the walls had set, whole hemp stems were used to create a platform to cover the roof with the insulation mix.

The thatching material was then fixed to the roof using hazel rod sways tied down with the protruding twine. As I didn't have enough hemp stems to completely thatch the roof, straw was used for the main part, with hemp for the ridge.

To finish the interior, hemp plaster was applied to the walls by hand and rubbed smooth. I then adapted a hempcrete mix for the floor that had a greater amount of lime in it and added sand to give it resistance to wear and tear as there was to be no other covering.

To protect the exterior of the walls a lime render was applied to the surface, again by smearing onto the hempcrete in order to keep the low-tech approach consistent throughout the project.

To give the whole thing a smart appearance two coats of lime paint were applied to the outside ,which will also help keep it weatherproof.

The girls insisted it was not finished or fit to play in until it had a door. So to be in keeping with the rest of the house a door was fashioned out of rough slabs, the hinges being the only hi-tech element involved.

Opposite; The completed wendy house.

The Future

One of the aims of this book is to encourage a future of hemp building. To point the way for future tests and experiments and to outline the possiblities that building ecologically using hemp can offer.

From a design point of view the basic aspects of a timber frame and an insulating warm layer of hempcrete can be put together as a home or shelter in an unending variety of ways, only limited by the imaginations of those designing them and the natural laws of physics.

The uncertain future regarding energy sources post-peak oil production will certainly focus minds on possible solutions to housing the global population in a more sustainable manner, and this system certainly has the answer to some of these questions.

By promoting a method of construction based primarily on material produced by the agricultural sector, we have to ensure that the production of hemp is a worthwhile one for the farmer. Future machinery developments may see the possibility of small-scale units enabling more localised processing of hemp which will be more appropriate, reducing the need for transporting the material large distances.

The method of producing hemp agriculturally speaking suits many part-time small organic farmers. The main activities of its husbandry only occupy a short period of time over the year, that of tillage and sowing, and then later the harvest. However, the cost of the seed is a factor of the whole plan that, even with the element of european government subsidies, still prevents many aspects of hemps potential from becoming possible.

The high cost of industrial hemp seed is due to the fact that in all "westernised countries" the growing of hemp is governed by the connection, by regulation, of indusrtial hemp to the psychoactive 'marijuana'.

This connection has fuelled the need for varieties of hemp with extremely low levels of THC (tetrahydracannibinol) to be developed with little regard to other qualities of the genetics. More importantly this means that farmers are required to purchase hybrid seed every year to be allowed permission to grow the crop.

This obsession with low THC content is absurd, to put it mildly. One might be correct in thinking that this is based entirely on the ignorance of the bureaucrats who drew up the regulations, if this is so, then here are the realities of the relationship between the cousins hemp & marijuana.

One can draw analogies between other closely related species used to provide food or clothing, such as beef and dairy cattle or members of the brassica family. Both can interbreed, but you will not get a decent steak from a milk cow or brussel sprouts from a spring cabbage.

Left; Past and present styles of timber frame houses

In the same way, if marijuana is sown in the manner of hemp, in closely planted rows, it will not be capable of growing fast enough to compete with the indigenous weeds. The alternative scenario of hiding a marijuana crop in amongst a hemp field is also impossible as the hemp crop will tower above the delicate 'dope plants' and prevent essential light from reaching them. Not only this but the whole aim of growing marijuana is to achieve a high level of THC. This is only present in large quantities in the resins exuded by the female flowers before they get fertilised with pollen from the male flowers. The whole focus of growers of high grade "sensimillia" is not to let the female flowers get fertilised so as to keep them in flowering mode and producing large resin covered buds.

In a field of hemp, as soon as the female plant starts to flower they are showered with pollen from the many male flowers present, thus preventing the production of any significant amount of the stimulant. In fact hemp fields are the bane of nearby marijuana farmers for this reason!

Despite the prohibition it is nowadays possible to obtain very strong cannabis buds in most towns or, the equipment and seeds to grow them hydroponically, on the internet. Anybody looking to procure or produce this commodity can do so without resorting to growing hemp.

The point is also made by newspaper reports of hemp industry projects that you can't get high on industrial hemp, this is true not only with hybrid seed but even produce from varieties with 20-30 times the levels of THC would still be regarded by fans of 'mary jane' to be unworthy of their attention. This rise in drug levels is hardly likely to happen throughout a seed crop of its own accord without careful selection of such mutations.

In a more productive scenario, hemp farmers would be able to save a portion of their crop for seed for the next season, as producers of wheat or barley might do. This would mean that hemp would be genuinely be able to become an economic reality as a product to replace many unsustainably produced materials.

In the meantime as regards usage in building, more comparison tests need to be carried out between hemp and other building systems using natural materials, to evaluate energy consumption during the production of materials and construction of such houses.

The only way in which this can be done realistically is to build houses using different systems and materials in the same location so that the local climate is the same.

Below; **Hemp being defoliated machanically prior to harvesting.**

The work being done in Montholier in France, to test two houses in the same housing estate; one of strawbale and the other of hempcrete, together with the tests done in Haverhill, England, are examples of this kind of experiment but a lot more needs to be done.

Universities are at a disadvantage nowadays as much of the funding for doing such tests comes from industry, the very industries that are producing materials the ecological builder is attempting to replace.

The producers of hemp products and many of the materials used with them are not yet in the position to fund a wide range of tests to fully evaluate the many facets of how a particular system or combination of materials work. In this way associations such as Construire en Chanvre are doing very valuable work in this field, by promoting the ideas and collecting and funding research.

At this point I would like to appeal to all those in the hemp and ecological building fields who hope to benefit from the growth in this sector to join this association to enable it to broaden its activities, as the work they have done so far is excellent and many parts of this book would have been impossible to write without it. I do understand that there may be vested interests who will resist this, but we are all learning.

Meanwhile as I write this book the price of oil is soaring and with it, all other sources of energy. Whether there are only a few years of oil left or many decades the "end of the world" cry has gone up many times before, without fruition thankfully. Those in the 'know' have predicted that London would be under 7 feet of horseshit by the 1930s, that oil would run out by the 1970s, and that electricity would be free anyday soon. All that seems to be true is that we constantly underestimate the potential of human ingenuity to be able to find a previously unforseen solution or reason why the disasters predicted will not occur.

This does not mean to say that we can all go out and buy fossil fuel burning S.U.V.s tomorrow and behave in an irre-

Left; Two houses built in Montholier, France. One (foreground) of hemp, the other (background) strawbale.

sponsible way.

I believe that of all the reasons to build in what is referred to as an ecologically sound method, the main one has to be that it has many definite benefits to our health.

To surround ourselves with materials that are emitting toxic fumes or that endanger those working to manufacture them is madness. To destroy environments and ravage forests to extract materials or fuel is equally foolish and for those who live in these regions horrific. Any way in which we can have a positive effect on the lives of our own families and those of others by how we spend our money must be worthwhile. We can at least vote with our wallets!

The economics of house building is another area that has room for improvement. Increasingly, many people are signing their lives away for the privilege of living under a roof. A situation that benefits no-one but the banks and financial institutions that proffer the contracts for such slavery. If the technology used in the building of ones home can be simplified to such an extent that not every task needs to be executed by expensive specialists or use costly materials then a situation could arise where it would be far more viable from a financial point of view to take a year off work to build ones home or at least help in the labour, than to spend the rest of your life paying for the privilege of having the job done by someone else. Group projects are evolving nowadays that encourage this approach, where neighbours all work on each others homes, an idea that it is hoped will also promote a better sense of community.

All this can sound like hopeless ideals of utopia to some ears, but my reply to this is always 'If you don't have a dream how you gonna have a dream come true!?

So that's it for now, happy hemp building.

All the best !

Steve Allin

Hemp

CZECH REPUBLIC

BENEU
Namesti Svateho Pavia 17,
27201 Kladno,
Cz.
Email;beneu@volny.cz

BRITAIN

HEMCORE Ltd
Latchmore Bank,Little hallingbury
Bishops stortford, Hertsford c
CM22 7PS
www.hemcore.com

CANADA

The Canadian Hemp Trade Alliance
1973 Halifax st.
Regins, SK. CANADA S4P 1T6
tel: (306) 757 4367
www.hemptrade.ca

FRANCE

La CHANVRIERE de L'AUBE
Rue Général De Gaulle,
10200 .
Bar-sur-Aube.
www.chanvre.com
(manufacturers of Chanvribat®)

CHANVRIERE du BELON
Les Kaolins BP3 29340 Reic sur Belon, Brittany.
tel; 0033 (0)298064534
www.chanvre-diffusion.com
(Organic materials)

TERRACHANVRE
(formally Kanabreizh)
Loriot Mine 22110 Trémargat
Tel; 0033(0)296365907
www.terrachanvre.com
(Organic materials)

CHANVRIBLOC®
Sarl Duport
Tel 0033(0)476812745
Fax 0033(0)476309794
email:duport.olivier@wanadoo.fr
(hempcrete blocks)

ISONAT®
Z.I. le Moulin 11 - BP 23
69470 Cours la Ville
fax 0033(0)474898889
contact@isonat.com
www.isonat.com
(insulation panels)

EFFIREAL
1,rue Nationale
49120 Chemille
tel;0033(0)241306019
Fax :0033(0)241303781
effireal@wanadoo.fr
www.effireal.ifrance.com
(Organic fibre lagging, matting, underlays & insulation panels)

HOLLAND

HEMPLAX BV
Postbus 142
9665 ZJ Oude Pekela
www.hempflax.com

INTERNATIONAL

HAMFHAUS
www.hamfhaus.com
(international hemp product suppliers)

HEMPTECH
www.hemptech.com
(information & publishing

THERMOHANF
Hock Vertriebs-Gmbh
Industriestrasse 7.D-76297
Stuttensee Spöck,Germany
(hemp fibre insulation)

ISOVER
Florapan Hempwool Insulation
Available through any Builder supplier of Isover products.
www.isover.com

IRELAND

Hemphab products
Kenmare
Co.Kerry
www.hempbuilding.com
(all types of hemp building materials)

Lime

BRITAIN

HYDRAULIC LIAS LIMES LTD.
Melmoth House, Abbey Close,
Sherbourne, Dorset. DT9 3LH
Tel: 00441 (0) 1458 223179

GERMANY

OTTERBEIN,
zement-u. kalkwerke GMbH Co
KG 36137 Grossenlüder-Müs
tel:00++(0)6648/68-0

IRELAND

THE TRADITIONAL LIME CO.
Rath, Shillelagh Rd.
Tullow , Co. Carlow.
www.traditionallime.co.uk

INTERNATIONAL

ST. ASTIER;
G.B stastier.co.uk
USA stastier.ca
etc.

BALTHAZARD & COTTE
ZAC de Valntin - BP 3011
25045 Besançon cedex
France.
tel : 0033(0)381474010
fax : oo33(0)381474019
eurovente.export@wanadoo.fr

VICAT
Bld des 3 vallons
38080 L'Isle d' Abeau
France
tel : 0033 (0) 474275907

Hemp builders

FRANCE

C.T.C.
Pratmeur
56310 Quistinc,
0033(0)297397044
goudet.laurent.ctc@wanadoo.fr

Charles Colgnard
le Tronchet
35380 Treffendel
Bretagne,
0033(0)299610130

IRELAND

Steve Allin
Environmental Design
Consultancy, Rusheens,
Kenmare. Co.Kerry.
www.hempbuilding.com
email ;hempbuilding@eircom.net

The Old Builders Company
tel; 086 8236451
www.oldbuilders.com
email ;hodt1@oldbuilders.com

SWITZERLAND

Pascal Pittet
Châtillens
www.pittet-artisan.ch
tel; 0041(0)219078244
email ;pascal@pittet-artisan.ch

Books

Architecture:

Eco
Elizabeth Wilhide
ISBN 1-903845-74-2

Environment and Services
Peter Burberry
ISBN 0 71341038 8

Rough Guide to Sustainability
Brian Edwards
ISBN 1 85946 1743

The Climatic Dwelling
Energy Research Group, U.C.D.
ISBN 1 873936 39 7

The Craft of Modular Post & Beam
James Mitchell
ISBN 0 88179 131 8

The Whole House Book
Pat Borer & Cindy Harris
ISBN 1898049211

Construire en Bois vol 1-& 2
Julius Natterer etc
ISBN 2 88074 258 7/250 1

Cannabis & Hemp:

Gargantua & Pantagruel
Rabelais
ISBN 1-84022-107-0

Green Gold & the Tree of Life
Chris Bennet etc
ISBN 09629872 2 0

Hemp Lifeline to the Future
Chris Conrad

Hemp Today
Ed Rosenthal
ISBN 0 932551 14 9

Marijuana Forbidden Medicine
Lester Grinspoon
ISBN 0 300 05435 1

Reefer Madness & Other Tales
Eric Schlosser
ISBN 0 141 01076 2

The Book of Grass
George Andrews
Published Pete Owen 1967

The Cultivation of Hemp
Dr Iván Bócsa & Michael Karus
ISBN 1 886874 03 4

The Emperor Wears No Clothes
Jack Herer
ISBN 1-878124-00-1

The Great Book of Hemp
Rowan Robinson
ISBN0 89281 541 8

Hemp Horizons
John Roulac
ISBN 0930031938

Materials & Skills:

Building with Lime
Stafford Holmes etc
ISBN 1 85339 384 3

La Chaux Naturelle
Julien Fouin
ISBN 9 782841 565603

Lime and Lime Mortars
Ad Cowper
ISBN 1 873394 29 2

Lime in Building
Jane Schofield
ISBN 0 9524341 2 1

Green Building Handbooks
Tom Woolley etc.
ISBN 0 419 22690 7

The Natural Paint Book
Lynn Edwards & Julia Lawless
ISBN 1 85626 432 7

Magazines & Periodicals

Construct Ireland
www.constructireland.ie

La Maison Ecologique
Le Petit Bourg
35630 Bazouges-sous-Hédé
www.la-maison-ecologique.com

Associations & Information

Construire en Chanvre
B.P. 6
F-89150 Saint-Valerien, France
www.
construction-chanvre.asso.fr

www.greenerbuildings.com
www.motherearthnews.com
www.resurgence.org
www.walnutbooks.com

Royal Istitute of British Architects
www.riba.org

Architects

Anthony Cohu RIBA
Borlin,Coomhola, Bantry, Co. Cork,
Ireland.
e mail: ajc46@iol.ie
photos ; pages 114,116,117

Ralph Carpenter
Modece Architects
88-89 St Johns St
Bury St Edmunds IP33 15Q
Suffolk U.K.
00441 284761141
www.modece.com
photos ; page 85

Techniques et Creations
Claude & Catherine Eichwald
5 rte de Weinbourg
67340 Weiterswiller, Alsace
techniques.creation@wanadoo.fr
photos ; pages 2,63,71,73

Peter Cowman
The Centre of Living Architecture
Carrick-on-shannon, Co. Leitrim
www.livingarchitecturecentre.com

Samuel Courgey
Arcanne
Rue de Vaulgrenant
39330 Pagnoz, France
www.arcanne.fr
photos ; pages 8,180

Bernard Menguy
25 rue Ile d'Arze
56000 Vannes.
France.
tel;0033(0)297405314
photos ; pages 60,112

Hemp building Research

ADEME
27, Rue Louis Vicat
75737 Paris cedex 15
tel oo33(0)147652000
www.ademe.fr

Building Research Establishment ltd.
Construction division,
Bucknalls Lane
Garston,Watford.WD25 9XX
England.
tel;0044(0)1923664200
www.bre.co.uk

Centre Scientifique et Techniques du Batiment
www.cstb.fr

Ecole Nationale des Travaux Public de L'Etat.
rue Maurice Ardin, F69518
Vaul-en-Velin cedex France.
tel;0033(0)472047285
www.entpe.fr

School of the Built Environment
The Queens University of Belfast
Belfast BT7 1NN
Northern Ireland. UK.
email: t.woolley@qub.ac.

The Hollies'

Centres of Alternative Building

Castletown roache
Enniskean,
Co Cork, Ireland.
00353(0) 23 47981
www.

Ecocentre de Perigord
Pegase-Creee
Froidefon
24450 St pierre de Frugie,
France.
tel; 00353(0)553525007

Embodied energy

	M J/kg
CEMENT	7.8
CONCRETE	1.3
GLASS	:
floated	15.9
laminated	16.3
toughened	26.2
GRAVEL	0.8
HARDWOOD	:
air dried & rough sawn	0.5
kiln dried & rough sawn	2.0
HEMP	7.1
INSULATION	:
cellulose	3.3
fibreglass	30.3
polystyrene	117.0
LIME	:
hydrated	6.5
GYPSUM PLASTER	4.5
PLASTERBOARD	6.1
POLYPROPYLENE	64.0
SHINGLES	9.0
SOFTWOOD	:
air dried & rough sawn	0.3
kiln dried & rough sawn	1.6
particle board	8.0
plywood	10.4
TERRACOTTA	15.5

Figures for embodied energy will vary from location to location depending on transport distances and types of energy and the systems used in their manufacture.

Hempcrete values

Co-efficeints :
λ [W/m.°C] = thermal conductivity
Cs [kJ/kg.°C] =specific heat capacity
Cvs [kJ/m.°C] = thermal volume
D_s [mm²/s] = thermal diffusity
Eff_s [J/m².s½.°C] = thermal effusity
ρ_{norm} [kg/m³] = mass/volume

Lightweight mix

λ [W/m.°C]	0.13
Cs [kJ/kg.°C]	1.4
Cvs [kJ/m.°C]	350
D_s [mm²/s]	0.17
Eff_s [J/m².s½.°C]	145
ρ_{norm} [kg/m³]	285

Wall mix

λ [W/m.°C]	0.17
Cs [kJ/kg.°C]	1.4
Cvs [kJ/m.°C]	565
D_s [mm²/s]	0.14
Eff_s [J/m².s½.°C]	215
ρ_{norm} [kg/m³]	445

Floor mix

λ [W/m.°C]	0.19
Cs [kJ/kg.°C]	1.4
Cvs [kJ/m.°C]	680
D_s [mm²/s]	0.13
Eff_s [J/m².s½.°C]	245
ρ_{norm} [kg/m³]	530

Plaster

λ [W/m.°C]	0.34
Cs [kJ/kg.°C]	1.4
Cvs[kJ/m.°C]	1310
D_s [mm²/s]	0.12
Eff_s [J/m².s½.°C]	460
ρ_{norm} [kg/m³]	990

All figures from Bétons de Chanvre published by Construire en Chanvre.

Material facts

1 hectare can produce 8-10 tonnes hemp
Material percentage of crop:
Fibre 15 - 20 %
Hurds 70 - 80 %
1 cubic metre of hemp hurds = 130 kg

A lime binder for hempcrete should have the following charecteristics;

LOW

Hydraulicity	< 0.17
Median diameter. **μ/n**	**<8**
Mass/volume kg/m3	**< 2600**
Thermic cvonductivity	**<0.12**

HIGH

Surface area cm or m2/g	**>5000**
Compression strength. M Pa	>1.00

The harder the setting strength of the lime the less 'free lime' it will contain.
the figures for St Asteir Naturel Hydraulic Limes are;

Nhl 2	35%
Nhl 3.5	25%
Nhl 5	15%

If sand is to be incorporated into the hempcrete or plaster it should be clean. The colour should be expected to alter the finish if it will be visible at the surface.

A

B

C

D

E

F

G

H

I

J

K

ALL PHOTOGRAPHS
© Steve Allin except:

© Martin Bankowski;
21.

© Ralph Carpenter;
85(right).

© Cannabioland;
23(bottom).

© Cloughreanne Lime;
39.

© Andrew Coffey;
14, 22 (bottom),24(bottom left,right) 28(bottom right), 172(bottom),179.

© Anthony Cohu;
114(right),116,163(bottom).

© Olivier Duport;
75(bottom right)

© Claude Eichwald;
63(bottom),73(bottom)

© Lucy Farmer;
124(top left & right).

© Laurent Goudet;
2,69(bottom right) 71(top & bottom) 115,133(top),141(bottom), 155(top), 169(bottom),170(bottom).

© Kevin Heapes;
113(bottom).

© Yvan Saint-Jours;
60, 112, 118(full page and inset), 119(top).

© La Chanvrière de l'Aube;
25(bottom) 34,69(bottom left),135(top)

© Loretta Leu;
6, 46(bottom) 77(right) 117.

© The Body Shop;
17(left).

© Ivan Watkins;
10(far left), 47(top right), 80,82,90,96,97(left), 105(bottom left), 106, 167.

© Tom Woolley;
74(top)

© Tim Yates BRE Ltd;
52, 54, 58

Cartoon on page 19
© Gilbert Shelton, by kind permission of Tony at Hassle Free Press.

Diagram of self erecting crane by kind permission of Don McColgan of Mantis crane hire.

Quotation from Gargantua & Pantagruel page 18. by kind permission of Dennis Hart at Wordsworth Editions Ltd

All illustrations © Aia Leu.

All diagrams by author.

Please do not presume that projects included in this book are open to public viewing because they have been photographed. Respect the privacy of the inhabitants.